Numerical Methods in Chemical Process Engineering Using Python

Hector Jorquera · Claudio A. Gelmi

Numerical Methods in Chemical Process Engineering Using Python

Tools for Modeling, Simulation, and Optimization

Springer

Hector Jorquera
Department of Chemical and Bioprocess Engineering
Pontificia Universidad Catolica de Chile
Santiago, Chile

Claudio A. Gelmi
Artificial Intelligence Center
Universidad San Sebastián
Santiago, Chile

ISBN 978-3-032-22957-1 ISBN 978-3-032-22958-8 (eBook)
https://doi.org/10.1007/978-3-032-22958-8

This Springer imprint is published by the registered company Springer Nature Switzerland AG
The registered company address is: Gewerbestrasse 11, 6330 Cham, Switzerland

To Francisca, Manuela, and Rocío, whose light fills my life.

To John S. Dahler, Rutherford Aris, H. Ted Davis, and Stephen Prager, who inspired me to strive and think deeply.

—Hector Jorquera

In memory of my late mentors,
Iván Solar and Babatunde (Tunde) Ogunnaike.
Their friendship, generosity, and unwavering passion for teaching shaped both this work and the person behind it.
With deep gratitude, I dedicate this book to their legacy.

—Claudio A. Gelmi

Preface

This university-level text has two main objectives: (i) to present concisely the most widely used numerical methods for solving the equations commonly encountered in Chemical Process Engineering (CPE). The emphasis is from the standpoint of someone who needs to solve such problems without delving into the details of numerical analysis itself (existence and uniqueness of solutions, convergence, etc.) and (ii) to show how Python and its scientific ecosystem (i.e., Matplotlib, NumPy, SciPy, pandas and Statsmodels) can be used to tackle the different categories of typical problems across the broad field of CPE. Python has become the most common language in introductory university programming courses globally and is now the de facto tool for research computing, from CERN's beam-dynamics simulations to NASA's prognostics packages. This explosive growth has been accompanied by a corresponding increase in university textbooks. Nevertheless, in our view there are still few books that link the most common numerical methods directly with the Python language, presenting the numerical technique and its implementation side by side, as this book does.

The first part of the text reviews the available numerical methods and shows how they can be implemented with Python libraries built-in functions, providing a perspective that complements the software's user manual and programming guides. Various techniques and algorithms are presented, highlighting which ones are best suited to each problem. Although "universal" algorithms rarely exist, we discuss the kinds of difficulties that may arise, how to diagnose them, and how to avoid them. Each chapter includes worked examples and a list of proposed problems; all have been tested in assessments for the courses "Applied Mathematics for Process Engineering" and "Computational Methods for Transport Phenomena," taught by the authors at the School of Engineering of the Pontificia Universidad Católica de Chile for more than 20 years.

The second part offers a deliberately selected set of solved problems, complete with code, results and figures. We have strived to include problems we consider most illustrative and engaging for students (parameter fitting with nonlinear equations and ordinary differential equations, confidence intervals, optimization of a bioreactor, etc.). As you will see, their level will challenge both undergraduate and graduate students. The problems are not limited to steady-state conditions; on the contrary, we have tried to include as many dynamic problems as possible. The general structure for solving each problem is straightforward: once the governing equations are derived or identified, the Python code is presented together with a brief explanation. Numerous graphics accompany the text, and in some cases the explanations go well beyond the immediate purpose of the problem, allowing students to explore related topics. We remind readers that programming—much like learning to integrate—is, in essence, an art. Other solutions may be completely valid; however, we have striven to offer the simplest and most direct ones.

All programs developed in this book are available at:

https://github.com/cagelmi/cpe-numerical-methods-python-book

You are invited to explore and modify them. The site will also host any errata that may arise (as with any written work, we are not immune to them).

We warmly thank Ediciones UC—particularly Patricia Corona, General Editor—for her support in allowing us to draw on selected elements of an earlier Spanish-language, MATLAB-focused edition, which has been substantially revised and expanded for the present Python-based text.

Finally, and no less importantly, we would like to express our deepest gratitude to our families for their understanding of the time taken to prepare this text.

Santiago, Chile
January 2026

Hector Jorquera, Ph.D.
Claudio A. Gelmi, Ph.D.

Competing Interests The authors have no competing interests to declare that are relevant to the content of this manuscript.

A View of Process Modeling and Simulation

A *model* can be defined as a set of hypotheses about the behavior of a real system, usually based on universal laws and principles, and expressed mathematically as equations—for example, Maxwell's equations for the electromagnetic field or the Navier–Stokes equations that describe the motion of a fluid under external forces. Table 1 proposes a classification of mathematical models.

Mathematical models are often used to design processes, for example, to scale laboratory results (where many parameters such as temperature are controlled) up to production scale, where more complex situations arise, such as temperature or humidity gradients, and where system behavior must be predicted before construction to optimize performance.

This is where the distinction between *modeling* and *simulation* comes in. *Modeling* usually refers to building a mathematical model of a real situation and demonstrating

Table 1 Classification of models according to their main characteristics

Type	Characteristics
Phenomenological	Equations are derived from scientific laws (physics, chemistry, biology, etc.)
Empirical	Equations are obtained from analyzing measured input and output data
Continuous	The model is valid for all values of a variable (time, space, etc.)
Discrete	The model applies only at certain time instants (every second, minute, hour, etc.)
Lumped-parameter	The model is described by a set of ordinary differential equations
Distributed	The model is described by a set of partial differential equations
Deterministic	Only deterministic variables are involved; each can take only one value in each situation
Stochastic	The model includes stochastic signals or random variables; certain variables have an associated probability distribution
Dynamic	Some model variables change with time
Static	All model variables are time-independent

that it can adequately reproduce the observed behavior of the system. *Simulation*, by contrast, is the application of an existing mathematical model to a new situation—for instance, forecasting tomorrow's weather, designing a heat exchanger under certain operating constraints, or simulating what happens if process conditions change.

As technology allows us to measure process characteristics (temperature, velocity, etc.) with ever-greater spatial and temporal resolution (e.g., nanomaterials), handling such information leads to increasingly complex quantitative models. Table 2 offers a classification of numerical problems in engineering based on their level of complexity.

Table 2 Typical engineering problems ranked by difficulty level

No. of stages	Information	Solution characteristics	Users	Example
1 or 2	All data are given	Numbers are substituted into a single equation	High-school students, technical personnel	Solve a quadratic equation
2 or 3	Data must be converted to consistent units	Numerical values are placed into equations that first need manipulation	First-year engineering students, technicians	Calibrate a thermometer from experimental results
4 or 5	Extra irrelevant or redundant data may exist	The quantitative relationship must be developed; the problem has an exact solution	Second-year engineering students, advanced technicians	Solve a linear-programming problem
Several	Data must be found	The problem has a unique solution; iterative methods and computer assistance may be needed	Second- and third-year engineering students	Size the diameter of a pipe for transporting a liquid
Many	Data must be selected and obtained from reliable sources	The problem has one solution, but a method is required; progress comes from learning from mistakes	Third-year engineering students	Calculate liquid–vapor equilibrium under non-ideal conditions
Many	Data may be contradictory or incomplete; accuracy must be estimated	Several solutions may exist, but one is optimal; approximations are needed, quick answers required, economics matter	Fourth-year engineering students, graduate students, design engineers	Calculate the optimal reflux ratio in a distillation column

(continued)

Table 2 (continued)

No. of stages	Information	Solution characteristics	Users	Example
Many routes	Additional data may be necessary	Problem may be unsolved; economic, environmental, and occupational-hygiene aspects are critical; starting points are not obvious	Graduate students, design engineers, R&D engineers	Cost estimation in a natural-product extraction process
Most routes fail	Required data may demand new measurement techniques	The final solution is often hard to recognize, and feasible solutions may not exist	Researchers, creative engineers	Development of a new vaccine

In summary, numerical modeling and simulation will continue to be invaluable tools for tackling new engineering problems. It is our hope that this book will help future professionals solve both classical and novel challenges.

Contents

About the Authors

Dr. Hector Jorquera is Professor in the Department of Chemical and Bioprocess Engineering, Pontificia Universidad Católica de Chile. He holds B.Sc and M.Sc degrees in Chemical Engineering from Universidad de Chile and a Ph.D. from the University of Minnesota, Minneapolis. Professor Jorquera has been teaching numerical methods to engineering students since 1992. He has been Visiting Professor in ENSBANA (Dijon, France), Imperial College (London, UK), University of Iowa (CGRER), Harvard University (ACMG) and National Center for Atmospheric Research (NCAR). He has published more than 60 papers in scientific journals. He is Author of the textbook *Introduction to Air Pollution Engineering* (in Spanish).

Dr. Claudio A. Gelmi earned a B.Sc. in Industrial and Chemical Engineering and a M.Sc. in Chemical Engineering from Pontificia Universidad Católica de Chile (UC) and a Ph.D. in Chemical Engineering from the University of Delaware (Newark, USA). He served as Assistant Professor in UC's Department of Chemical and Bioprocess Engineering and held leadership roles as Associate Dean for Engineering Education in the School of Engineering and Director of the Center for Teaching Development within the Academic Vice-Rector's Office. He has authored more than 20 research articles and two textbooks, among other publications. He has led data-driven initiatives as an independent consultant and taught artificial intelligence in UC executive programs. He currently leads the Artificial Intelligence Center at Universidad San Sebastián (Chile).

Part I
Foundations of Numerical Methods

Chapter 1
Systems of Linear Equations

The numerical solution of systems of linear equations can be divided into:

(a) Direct Method, if the number of steps required to solve the equations is finite. For example, the solution of the linear system $\mathbf{A}\cdot\mathbf{x} = \mathbf{b}$ when the inverse of matrix $\mathbf{A}$ exists.
(b) Iterative Method, if an infinite number of steps is required to solve the equations exactly. For example, solving a linear system of equations by the Jacobi method.

1.1 Direct Solution Methods

A direct method is an algorithm with a finite and predefined number of steps, at the end of which the solution is obtained.

1.1.1 Gauss-Jordan Elimination

This algorithm consists of performing a series of transformations on the original linear system until an upper triangular system is obtained. Suppose our starting point is the linear system

$$\mathbf{A} \cdot \mathbf{x} = \mathbf{b} \tag{1.1}$$

A sequence of operations is generated until the original system of equations is transformed into a system with an upper triangular matrix U:

$$\mathbf{A}^{(\mathbf{n})} \cdot \mathbf{x} = \mathbf{U} \cdot \mathbf{x} = \mathbf{b}^{(\mathbf{n})} \tag{1.2}$$

H. Jorquera and C. A. Gelmi, *Numerical Methods in Chemical Process Engineering Using Python*, https://doi.org/10.1007/978-3-032-22958-8_1

The solution to this equation is the same as that of Eq. 1.1.

To obtain the upper triangular structure, the elements below the main diagonal are eliminated, making them zero through row operations (additions and subtractions) on matrix A. At the end of the process (referred to as pivoting), the following matrix structure is obtained (Jorquera and Gelmi, 2014)

$$A^{(n)} = \begin{bmatrix} a_{11} & a_{12} & . & . & . & . & a_{1n} \\ 0 & a_{22} & a_{23} & . & . & . & a_{2n} \\ 0 & 0 & a_{33} & a_{34}. & . & . & a_{3n} \\ \dots & \dots & \dots & \dots & \dots & \dots & \dots \\ 0 & 0 & 0 & 0 & a_{n-2,n-2} & & \\ 0 & 0 & 0 & 0 & 0 & a_{n-1,n-1} & \\ 0 & 0 & 0 & 0 & 0 & 0 & a_{nn} \end{bmatrix} \tag{1.3}$$

This yields the system of equations $\mathbf{U} \cdot \mathbf{x} = \mathbf{y}$, which can be solved by back substitution to obtain the solution $\mathbf{x}$.

Example 1.1 Gauss-Jordan elimination

Use Python to perform Gauss-Jordan elimination of the following system of equations

$$\mathbf{A} \cdot \mathbf{x} = \mathbf{b},$$

$$A = \begin{bmatrix} 2 & 1 & -3 \\ -1 & 3 & 2 \\ 3 & 1 & -3 \end{bmatrix}; \quad b = \begin{bmatrix} -1 \\ 12 \\ 0 \end{bmatrix}$$

Solution: Pivoting is performed on matrix **A** together with the vector **b** to produce the modified system of equations. For this purpose, an augmented matrix **M** = [**A** **b**] is used. The following Python script solves the proposed problem:

```
import numpy as np

# -------------------------------------------------
# This script solves Example 1.1
# -------------------------------------------------

# Input of matrix A and right-hand-side vector b
A = np.array([[2, 1, -3],
              [-1, 3,  2],
              [3, 1, -3]], dtype=float)

b = np.array([-1, 12, 0], dtype=float)

# Initialize solution vector
x = np.zeros_like(b)

# -------------------------------------------------
# Definition of augmented matrix M and Gauss-Jordan elimination
# -------------------------------------------------
M = np.hstack((A, b.reshape(-1, 1)))

# Normalize row 1 using the first pivot
M[0, :] = M[0, :] / M[0, 0]

# Generate a zero in position (2,1)
M[1, :] = M[1, :] - M[0, :] * M[1, 0]

# Generate a zero in position (3,1)
M[2, :] = M[2, :] - M[0, :] * M[2, 0]

# Normalize row 2 using the second pivot
M[1, :] = M[1, :] / M[1, 1]

# Generate a zero in position (3,2)
M[2, :] = M[2, :] - M[1, :] * M[2, 1]

# -------------------------------------------------
# Equation solution stage using back substitution
# -------------------------------------------------
x[2] = M[2, 3] / M[2, 2]
x[1] = M[1, 3] - M[1, 2] * x[2]
x[0] = M[0, 3] - M[0, 2] * x[2] - M[0, 1] * x[1]

# -------------------------------------------------
# Solution verification
# -------------------------------------------------
r = A @ x - b

print("Augmented matrix:")
print(M)

print("Solution x:")
print(x)

print("\nResidual r = A x - b:")
print(r)
```

The result of applying this script is:

$$\boldsymbol{M} = \begin{bmatrix} 1 & 0.5 & -1.5 & -0.5 \\ 0 & 1 & 0.1428 & 3.2857 \\ 0 & 0 & 1.5714 & 3.1428 \end{bmatrix}; \quad \boldsymbol{x} = \begin{bmatrix} 1 \\ 3 \\ 2 \end{bmatrix}; \quad \boldsymbol{r} = \begin{bmatrix} 0 \\ 0 \\ 0 \end{bmatrix}$$

Notes:

1. The method as described here (Gauss-Jordan elimination) fails if any pivot (the diagonal elements of $\mathbf{A}^{(n)}$) becomes zero or is very small due to rounding errors (e.g., subtraction of quantities of similar magnitude).

2. In practice, a pivoting strategy is necessary to select the "best" pivot to minimize numerical issues. The best pivot is typically the element with the largest absolute value in the column that remains below the main diagonal during forward elimination.
3. In the case of positive definite matrices or matrices with diagonal dominance,[1] pivoting always works and there are no issues with encountering pivots equal to zero.

1.1.2 Case of Tridiagonal Matrices

A simple example of Gauss-Jordan elimination is the case of a tridiagonal matrix, of the form

$$\mathbf{A} = \begin{bmatrix} \alpha_1 & \gamma_1 & & & & \\ \beta_1 & \alpha_2 & \gamma_2 & & & \\ & \cdot & \cdot & \cdot & & \\ & & \cdot & \cdot & \cdot & \\ & & & \cdot & \cdot & \cdot \\ & & & & \cdot & \gamma_{n-1} \\ & & & & \beta_{n-1} & \alpha_n \end{bmatrix} \tag{1.4}$$

If the diagonal dominance condition is satisfied for the matrix: $|\alpha_i| > |\gamma_i| + |\beta_{i-1}|$, for all i, then the Gauss-Jordan elimination process is described by the Thomas algorithm, which starts with the forward pivoting stage

$$\begin{aligned} &\alpha_1^* = \alpha_1;\ b_1^* = b_1 \\ &m_k = \frac{\beta_{k-1}}{\alpha_{k-1}^*};\ \alpha_k^* = \alpha_k - m_k \cdot \gamma_{k-1};\ k = 2, \ldots, n \\ &b_k^* = b_k - m_k \cdot b_{k-1}^*;\ k = 2, \ldots, n \end{aligned} \tag{1.5}$$

Followed by a backward substitution stage:

$$x_n = \frac{b_n^*}{\alpha_n^*};\quad x_k = \frac{b_k - \gamma_k \cdot x_{k+1}}{\alpha_k};\quad k = n-1,\ n-2, \ldots, 1 \tag{1.6}$$

[1] A matrix $\mathbf{A}$ is said to be diagonally dominant if: $|a_{ii}| > \sum_{j \neq i} |a_{ij}|$ for i = 1, 2, … n.

Notes:

1. This algorithm not only avoids unnecessary arithmetic operations on zero elements of the matrix, but also reduces the required storage from $O(n^2)$ to approximately $O(5 \cdot n)$, since matrix $\mathbf{A}$ is stored as three vectors $\boldsymbol{\alpha}$, $\boldsymbol{\beta}$, and $\boldsymbol{\gamma}$. This algorithm is implemented in the next Example.
2. This approach can be implemented efficiently when the original matrix possesses a regular structure. This tridiagonal matrix frequently arises when modeling mass balances in separation processes such as distillation (Wang & Henke, 1966), absorption, solvent extraction (Hanna & Sandall, 1995); it also appears in boundary value problems (ordinary or partial differential equations) when finite-difference methods are used (Chaps. 4 and 5).

Example 1.2 Solution of a tridiagonal system

Aniline is removed from an aqueous stream using toluene as the extracting solvent; the water flow (F_1) and the toluene flow (F_2) are constant (on a solvent-free basis). The separation unit is a countercurrent column with 10 extraction stages, as shown in Fig. 1.1, with the following notation:

x_i = kg of aniline in the organic phase/kg of toluene in the organic phase.
y_i = kg of aniline in the aqueous phase/kg of water in the aqueous phase.

The subscripts "i" correspond to the compositions of the streams exiting the ith extraction stage. The mass balance equations for each stage "i" can be written as follows:

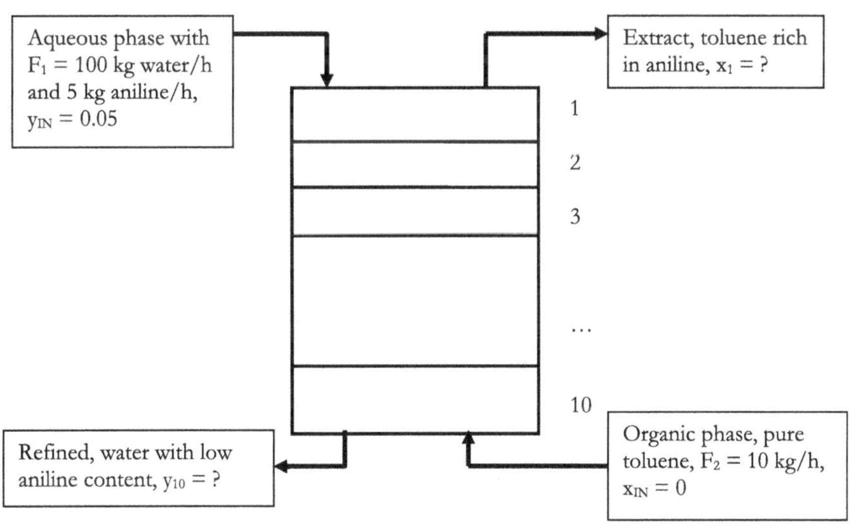

Fig. 1.1 Diagram of the liquid–liquid extraction of Example 1.2

$$y_{i-1} - \left(1 + \frac{F_2}{F_1} \cdot K\right) \cdot y_i + \left(\frac{F_2}{F_1} \cdot K\right) \cdot y_{i+1} = 0; \quad K = \frac{x_i}{y_i}; \; i = 2, 3, \ldots, 9$$

where K is the solute distribution coefficient between the two phases, assumed constant across all stages. For stages 1 and 10, the following equations apply:

$$-\left(1 + \frac{F_2}{F_1} \cdot K\right) \cdot y_1 + \left(\frac{F_2}{F_1} \cdot K\right) \cdot y_2 = -y_{IN};$$

$$y_9 - \left(1 + \frac{F_2}{F_1} \cdot K\right) \cdot y_{10} = -\left(\frac{F_2}{F_1}\right) \cdot x_{IN}$$

Build a Python macro that solves this system of equations and plots the concentrations of aniline in both phases as a function of the number of stages, assuming $K = 4$ as a constant.

The following Python script solves the system of equations of Example 1.2 using Thomas algorithm for tridiagonal matrices, and the results are presented in Fig. 1.2.

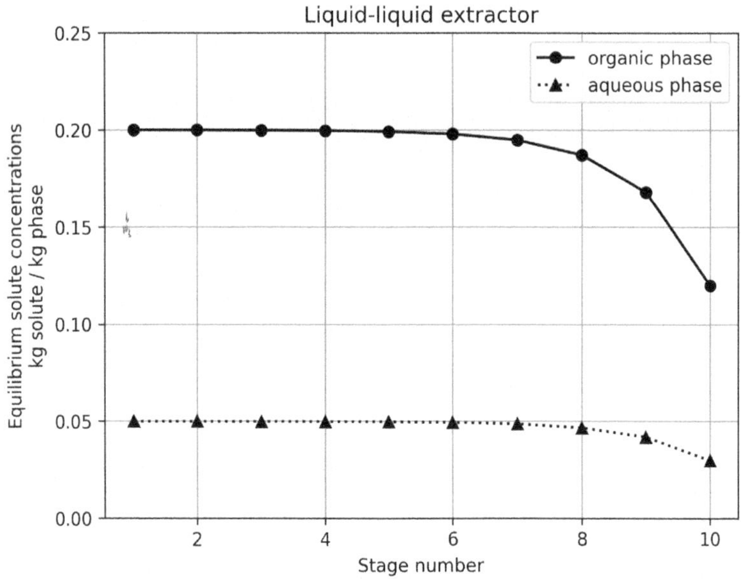

Fig. 1.2 Solution of the tridiagonal system of equations

```python
import numpy as np
import matplotlib.pyplot as plt

# -------------------------------------------------
# This script solves Example 1.2
# -------------------------------------------------

F1 = 100
F2 = 10
x_in = 0
y_in = 0.05
K = 4

# Common factor used in the mass balances
alpha = F2 * K / F1

# Construction of the right-hand-side vector of the linear system
b = np.zeros(10)
b[0] = -y_in
b[-1] = -F2 * x_in / F1

# Construction of the three diagonals of the linear system
e = np.ones(10)                  # lower diagonal
f = -(1 + alpha) * e             # main diagonal
g = alpha * e                    # upper diagonal

# -------------------------------------------------
# Tridiagonal system solver (Thomas algorithm)
# -------------------------------------------------
def tridisolve(a, b, c, d):
    """
    Solves a tridiagonal linear system using the Thomas algorithm
    a: lower diagonal
    b: main diagonal
    c: upper diagonal
    d: right-hand-side vector
    """
    n = len(d)
    cp = np.zeros(n)
    dp = np.zeros(n)

    cp[0] = c[0] / b[0]
    dp[0] = d[0] / b[0]

    for i in range(1, n):
        denom = b[i] - a[i] * cp[i - 1]
        cp[i] = c[i] / denom if i < n - 1 else 0.0
        dp[i] = (d[i] - a[i] * dp[i - 1]) / denom

    x = np.zeros(n)
    x[-1] = dp[-1]
    for i in reversed(range(n - 1)):
        x[i] = dp[i] - cp[i] * x[i + 1]

    return x

# -------------------------------------------------
# Solution for concentrations {y_i}
# -------------------------------------------------
y = tridisolve(e, f, g, b)

# Solution for concentrations {x_i}
```

```
x = K * y

# ---------------------------------------------------
# Plot of equilibrium concentrations in each phase
# ---------------------------------------------------
stage = np.arange(1, 11)

plt.figure()
plt.plot(stage, x, 'k-o', label='organic phase')
plt.plot(stage, y, 'k:^', label='aqueous phase')
plt.title('Liquid-liquid extractor')
plt.xlabel('Stage number')
plt.ylabel('Equilibrium solute concentrations\nkg solute / kg phase')
plt.ylim(0, 0.25)
plt.legend()
plt.grid(True)

# Save figure
filename = 'Figure_1_2.png'
plt.savefig(filename, dpi=300, bbox_inches='tight')
plt.show()

# Show numerical results

print(" x:")
print(x)

print(" y:")
print(y)
```

Note that five equilibrium stages are sufficient for the concentrations of the solute to no longer change, indicating that no additional solute can be extracted in the organic phase; the aqueous-phase outlet has a concentration $y_{10} = 0.03$ kg/kg. The reader is encouraged to construct a Python function that solves the general case of N equilibrium stages for a liquid–liquid extractor, allowing for an initial solute content in the solvent ($x_{IN} > 0$).

1.1.3 Number of Operations Required

Computational work, in terms of multiplications and additions, is an important measure of the algorithm's efficiency. For the Gauss-Jordan elimination in a system of n equations, the number of operations is given by

- Elimination phase: $n^3/3 + O(n^2)$
- Back substitution phase: $n^2/2 + O(n)$

In other words, the number of operations grows with the cube of the system dimension. If we consider that today's intensive applications involve n in the order of one million elements, we can appreciate that the highest complexity quickly leads to very substantial computational costs for this class of algorithms. We will revisit this issue when we examine iterative methods.

1.1.4 Direct Methods Implemented in Python

In Python, we have several options to solve systems of linear equations:

(a) An obvious alternative is to use the function that computes the inverse of a matrix, in which case the solution is given by **x** = np.linalg.inv(**A**)@**b**. However, this approach is limited to square matrices that possess an inverse and are well-conditioned (see Sect. 1.3); in addition, it is not the most efficient method when the matrix has certain structures such as being triangular, symmetric, having few diagonal bands and zeros elsewhere, etc.
(b) Python routine np.linalg.solve(**A**, **b**) corresponds to solving the basic equation **A** @ **x** = **b**.
(c) LU factorization enables efficient solution of the system Eq. (1.1) when the solution **x** must be evaluated for multiple right-hand sides **b**. In Python this is done using: P, L, U = lu(A).

However, all these functions rely on Gauss-Jordan elimination, which is known to become costly as the number of equations increases. Consequently, it is necessary to consider alternative solution schemes for Eq. (1.1) that do not involve forming the inverse of matrix A.

Example 1.3 Comparison of solution methods

Solve the system of equations from Example 1.1 using Python's np.linalg.inv, np.linalg.solve, or lu function followed by solve_triangular.

The following macro applies the three methodologies to solve the system of equations; all three methods yield the same solution, and the residuals are zero. In the case of LU factorization, the routine returns a permutation matrix **P** such that **P**@**A** = **L**@**U**; this matrix **P** represents the row interchanges performed in the partial pivoting strategy to prevent zeros (or near zeros) on the diagonal.

```
import numpy as np
from scipy.linalg import lu

# --------------------------------------------------
# This script solves Example 1.3
# --------------------------------------------------

# Input of matrix A and right-hand-side vector b
A = np.array([[2, 1, -3],
              [-1, 3,  2],
              [3, 1, -3]], dtype=float)

b = np.array([-1, 12, 0], dtype=float)

# --------------------------------------------------
# Solution using the inverse of matrix A
# --------------------------------------------------
x1 = np.linalg.inv(A) @ b
r1 = A @ x1 - b

# --------------------------------------------------
# Solution using the backslash operator equivalent (solve)
# --------------------------------------------------
x2 = np.linalg.solve(A, b)
r2 = A @ x2 - b

# --------------------------------------------------
# Solution using LU factorization
# --------------------------------------------------
P, L, U = lu(A)
b3 = P @ b
y3 = np.linalg.solve(L, b3)
x3 = np.linalg.solve(U, y3)
r3 = A @ x3 - b

# --------------------------------------------------
# Print results
# --------------------------------------------------
print("Solution using inverse:")
print(x1)
print("Residual:", r1)

print("\nSolution using solve:")
print(x2)
print("Residual:", r2)

print("\nSolution using LU factorization:")
print(x3)
print("Residual:", r3)
```

1.2 Iterative Methods

There exists a wide variety of iterative methods applied to systems of linear equations. These methods are often used in solving linear systems with many variables ($n >> 1$), which are costly to solve via Gaussian elimination. Most of these methods are stationary iterations of the form (Axelson, 1994):

$$\boldsymbol{x}^{(k+1)} = \boldsymbol{M} \cdot \boldsymbol{x}^{(k)} + \boldsymbol{g}; \quad \boldsymbol{k} = \mathbf{1}, \mathbf{2}, \ldots \tag{1.7}$$

Usually, the matrix **M** and the vector **g** are independent of the iteration index k. A sufficient condition for the convergence of the iteration Eq. (1.7) is that the norm of the matrix **M** is less than 1. A necessary and sufficient condition for convergence is that

$$\rho(\boldsymbol{M}) < 1; \quad \rho(\boldsymbol{M}) \equiv \max|\lambda_i|; \quad i = 1, 2, \ldots, n \tag{1.8}$$

where $\rho(\mathbf{M})$ is the spectral radius of **M** and corresponds to the largest absolute value of the eigenvalues of M. In Python this would be computed as ρ (**M**) = np.max(np.abs(np.linalg.eigvals(**M**))).

The iteration given by Eq. (1.7) is obtained from the original equation $\mathbf{A} \cdot \mathbf{x} = \mathbf{b}$ by a decomposition of the original matrix **A**. We can decompose matrix A in the following manner:

$$\begin{aligned} &\boldsymbol{A} = \boldsymbol{D} + \boldsymbol{L} + \boldsymbol{U}; \quad \boldsymbol{D} = \boldsymbol{diag}(\boldsymbol{a_{ii}}) \\ &\boldsymbol{L} = \begin{Bmatrix} a_{ij} \ if & i > j \\ 0 \ if & i \leq j \end{Bmatrix}; \quad U = \begin{Bmatrix} a_{ij} \ if & i < j \\ 0 \ if & i \geq j \end{Bmatrix} \end{aligned} \tag{1.9}$$

Note that in Python we can generate the decomposition in the following way: **D** = np.diag (np.diag(**A**)); **L** = tril(**A**, −1); **U** = triu(**A**, 1). From this decomposition it is possible to obtain several schemes that have been frequently used as iterative methods (Gerald, 1998; Nakamura, 1991), and which are described below.

1.2.1 Jacobi Method (Simultaneous Displacements)

In this method, the decomposition Eq. (1.9) is applied, and the diagonal of the original matrix is placed on the left-hand side of the original equation, so that

$$\begin{aligned} &\boldsymbol{D} \cdot \boldsymbol{x}^{(k+1)} = -(\boldsymbol{L} + \boldsymbol{U}) \cdot \boldsymbol{x}^{(k)} + \boldsymbol{b}; \\ &\boldsymbol{x}^{(k+1)} = \left\{-\boldsymbol{D}^{-1} \cdot (\boldsymbol{L} + \boldsymbol{U})\right\} \cdot \boldsymbol{x}^{(k)} + \boldsymbol{D}^{-1} \cdot \boldsymbol{b} \end{aligned} \tag{1.10}$$

Therefore, in each iteration the new components of the solution vector are computed using information from the components of the previous stage, so that in Eq. (1.10) the i-th component of the updated solution vector is given by:

$$x_i^{(k+1)} = \frac{1}{a_{ii}} \cdot \left\{ -\sum_{j=1, j \neq i}^{n} a_{ij} x_j^{(k)} + b_i \right\}; \quad i = 1, 2, \ldots, n \tag{1.11}$$

1.2.2 Gauss–Seidel Method (Successive Displacements)

In this variant, only the upper diagonal of matrix A is moved to the right-hand side of the original equation, yielding the following expression:

$$(\boldsymbol{D}+\boldsymbol{L})\cdot\boldsymbol{x}^{(k+1)} = -\boldsymbol{U}\cdot\boldsymbol{x}^{(k)}+\boldsymbol{b};$$
$$\boldsymbol{x}^{(k+1)} = \left\{-(\boldsymbol{D}+\boldsymbol{L})^{-1}\cdot\boldsymbol{U}\right\}\cdot\boldsymbol{x}^{(k)}+(\boldsymbol{D}+\boldsymbol{L})^{-1}\cdot\boldsymbol{b} \tag{1.12}$$

In this method, the i-th component of the solution vector at stage $(k+1)$ is updated using information from the components of the vector $\mathbf{x}^{(k)}$ and the newly computed components of $\mathbf{x}^{(k+1)}$ up to component $i-1$.

$$x_i^{(k+1)} = \frac{1}{a_{ii}}\left\{-\sum_{j=1}^{i-1} a_{ij}x_j^{(k+1)} - \sum_{j=i+1}^{n} a_{ij}x_j^{(k)} + b_i\right\} \tag{1.13}$$

In other words, more up-to-date information is utilized than in the Jacobi method.

1.2.3 Successive Relaxation Method

The greater speed of convergence of the Gauss–Seidel method (with respect to the Jacobi method) is due to updating all available information of the vector $\mathbf{x}^{(k+1)}$ to compute the remaining components of that vector, and this is achieved by including the **L** part of the original matrix **A** on the left-hand side of Eq. 1.10.

In the successive relaxation method, the possibility of accelerating the convergence of Gauss–Seidel iterations using a relaxation parameter ω is considered, so that an iteration update is performed in the form:

$$\begin{aligned}(\boldsymbol{D}+\omega\cdot\boldsymbol{L})\cdot\boldsymbol{x}^{(k+1)} &= -\{(\omega-1)\cdot\boldsymbol{D}+\omega\cdot\boldsymbol{U}\}\cdot\boldsymbol{x}^{(k)}+\omega\cdot\boldsymbol{b}\\ \rightarrow \boldsymbol{x}^{(k+1)} &= (\boldsymbol{D}/\omega+\boldsymbol{L})^{-1}\cdot\left[\{(\mathbf{1}/\omega-1)\cdot\boldsymbol{D}-\boldsymbol{U}\}\cdot\boldsymbol{x}^{(k)}+\boldsymbol{b}\right]\end{aligned} \tag{1.14}$$

The parameter ω is greater than 0 and less than 2. If ω is greater than 1.0, we refer to over-relaxation, whereas if $0 < \omega < 1$, the method is called under-relaxation. There exist techniques to estimate the optimal value of the parameter ω for a given situation (Axelson, 1994), including adjusting its value during the iteration (adaptive schemes), which essentially consist of minimizing the spectral radius (ρ) of the matrix $\mathbf{M} = (\mathbf{D}/\omega + \mathbf{L})^{-1}\cdot((1/\omega-1)\cdot\mathbf{D} - \mathbf{U})$ that appears in Eq. (1.14).

Notes:

1. The Jacobi and Gauss–Seidel methods are useful only when the matrix is diagonally dominant, which guarantees the convergence of both methods.
2. Typically, in increasing order of convergence speed, Jacobi < Gauss–Seidel < Relaxations, since in the latter case one can choose ω to minimize ρ (**M**).
3. It can be shown that if $0 < \omega < 2$ and **A** is positive definite, the successive relaxation method converges (Sewell, 1988).
4. It can be shown that if $0 < \omega < 1$ and **A** is diagonally dominant, the successive relaxation method converges (Sewell, 1988).

Example 1.4 Comparison of Iterative Methods

For the following systems of linear equations, compute the number of iterations required to achieve a residual value smaller than 10^{-5} in the norm of the residual $\mathbf{r}^{(k)} = \mathbf{A}\cdot\mathbf{x}^{(k)} - \mathbf{b}$ using the following methods: (i) Jacobi, (ii) Gauss–Seidel, (iii) relaxations applied to Gauss–Seidel, choosing ω in order to minimize ρ (**M**) in the Gauss–Seidel method.

$$\text{(a)}\quad \boldsymbol{A} = \begin{bmatrix} 6 & -3 & 1 \\ 1 & -7 & 1 \\ 2 & 1 & -8 \end{bmatrix};\quad \boldsymbol{b} = \begin{bmatrix} 11 \\ 10 \\ -15 \end{bmatrix};$$

$$\text{(b)}\quad \boldsymbol{A} = \begin{bmatrix} 1 & 4 & 7 \\ 2 & 5 & 8 \\ 3 & 6 & 11 \end{bmatrix};\quad \boldsymbol{b} = \begin{bmatrix} 1 \\ 1 \\ 1 \end{bmatrix}$$

$$\text{(c)}\quad \boldsymbol{A} = \begin{bmatrix} 4 & -1 & 0 \\ -1 & 4 & -1 \\ 0 & -1 & 4 \end{bmatrix};\quad \boldsymbol{b} = \begin{bmatrix} 6 \\ 4 \\ 6 \end{bmatrix}$$

Applying the various iterative schemes yields the results in Table 1.1.

Therefore, a relevant conclusion is that all iterative methods of the general form Eq. (1.7) must undergo a preliminary check to ensure that the convergence conditions are satisfied before attempting to solve them by any iterative method.

Below, we present a Python code listing for iterative resolution using the Jacobi method and its application to system a); the reader can easily modify the code to implement the Gauss–Seidel method or the successive-relaxation method and check the results in Table 1.1.

Table 1.1 Results for Example 1.4

Linear system	Solution of the equations	N° of iterations needed for *np.linalg.norm* ($\mathbf{A}\cdot\mathbf{x} - \mathbf{b}$) $< 10^{-5}$		
		Jacobi	Gauss–Seidel	SOR[§]
a	$[1\ -1\ 2]^T$	11	7	8
b	$[-1/3\ 1/3\ 0]^T$	Does not converge	Does not converge	Does not converge
c	$[2\ 2\ 2]^T$	14	8	6

[§]Computed using the value of ω that minimizes ρ(**M**) in Eq. 1.14

Note that case (b) does not converge because the matrix is neither diagonally dominant nor can the rows be reordered to achieve diagonality. For cases (a) and (c), both matrices are diagonally dominant, but only (c) is positive definite (all its eigenvalues are strictly positive); hence the relaxation method attains its greatest implementation efficiency

```
import numpy as np

def jacobi(A, b, tol, maxit):
    """
    Function that solves A*x = b using the Jacobi method
    """

    # Diagonal matrix
    D = np.diag(np.diag(A))

    # Strictly lower triangular matrix
    L = np.tril(A, -1)

    # Strictly upper triangular matrix
    U = np.triu(A, 1)

    # Iteration matrix
    M = -np.linalg.inv(D) @ (L + U)

    # Check that the eigenvalues satisfy the convergence condition
    spectral_radius = np.max(np.abs(np.linalg.eigvals(M)))
    if spectral_radius >= 1.0:
        raise ValueError("Iteration matrix does not guarantee convergence")

    # Constant vector
    g = np.linalg.solve(D, b)

    # Initial guess
    X = np.zeros_like(b, dtype=float)

    it = 0
    err = np.linalg.norm(b)

    # Jacobi iteration loop
    while err > tol and it < maxit:
        X = M @ X + g
        err = np.linalg.norm(A @ X - b)
        it += 1

    return X, it, err
import numpy as np

# Coefficient matrix
A = np.array([[6, -3,  1],
              [1, -7,  1],
              [2,  1, -8]], dtype=float)

# Right-hand-side vector
b = np.array([11, 10, -15], dtype=float)

# Tolerance and maximum number of iterations
tol = 1e-5
maxit = 100

# Call to the Jacobi method
X, it, err = jacobi(A, b, tol, maxit)

print("Solution X:")
print(X)

print("\nNumber of iterations:")
print(it)

print("\nFinal error (residual norm):")
print(err)
```

1.2.4 Error Estimation in Iterative Methods

Let $\mathbf{x}^*$ be the solution of the generic Eq. (1.7); it can be shown that a bound for the absolute error after the k-th iteration stage is given by

$$\left\| x^{(k+1)} - x^* \right\| \le \frac{\|M^k\|}{1 - \|M\|} \left\| x^{(2)} - x^{(1)} \right\| \tag{1.15}$$

Therefore, for the present method to converge, it is necessary that the norm of the matrix **M** be < 1.0. This is equivalent to requiring that the spectral radius of **M**, defined in Eq. (1.8), be less than 1.0.

Notwithstanding the foregoing, if the norm of **M** is greater than 1.0, it is possible to solve Eq. (1.7) indirectly using the residuals $\mathbf{r}^{(j)}$ of that iteration, via conjugate gradient methods (Axelson, 1994; Sewell, 1988), which are more complex to implement and whose performance depends strongly on the spectral structure of the eigenvalues of matrix **M**, making its performance difficult to predict.

In the case that we must solve iteratively a scenario such as the one just described, one would need to test the iterative methods provided by Python, described in the following section.

1.2.5 Iterative Methods Implemented in Python

The following iterative methods are implemented in the Python environment:

- Generalized Minimal Residual (GMRES) method: scipy.sparse.linalg.gmres, for general, non-symmetric systems.
- Conjugate Gradient (CG) method: scipy.sparse.linalg.cg, for symmetric, positive definite systems (SPD).
- Minimal Residual (MinRes) method: scipy.sparse.linalg.minres, for symmetric (non-SPD) systems.
- Bi-Conjugate Gradient (BiCG) method: scipy.sparse.linalg.bicg, for non-symmetric systems.
- Bi-Conjugate Gradient Stabilized (BiCGSTAB) method: scipy.sparse.linalg.bicgstab, for non-symmetric systems.
- Quasi-Minimal Residual (QMR) method: scipy.sparse.linalg.qmr, for non-symmetric systems.

Most of these algorithms benefit from pre-conditioning matrix **A**. Some of these algorithms are: Jacobi (diagonal scaling), Incomplete LU (ILU/IC), Algebraic Multi-grid (AMG), Incomplete Cholesky (ICC), Sparse Approximate Inverse (SPAI) and Block preconditioners for block-structured systems. The Python packages PyAMG and scikit-sparse support the pre-conditioning steps. For more details on specific applications and the original references for the methods, the reader is encouraged to solve the sparse matrix problems in Sect. 1.5.

1.3 Error Analysis

Let us consider the possibility that there are errors in the problem input data and, therefore, that it is necessary to estimate the magnitude of the error in the solution of the equations. For these purposes, the inverse matrix $\mathbf{A}^{-1}$ is fundamental in determining the solution error.

When small errors in the coefficient matrix **A** produce large changes in the solution, the system of equations is said to be ill-conditioned. This is typically due to the rows of **A** being nearly linearly dependent.

Consider the case when there is an error in the right-hand side of the equations, that is,

$$\boldsymbol{A} \cdot \boldsymbol{x} = \boldsymbol{b} + \boldsymbol{\delta b} \tag{1.16}$$

Whose solution is $\mathbf{x} + \boldsymbol{\delta}\mathbf{x}$. Then it follows that

$$\boldsymbol{\delta x} = \boldsymbol{A}^{-1} \cdot \boldsymbol{\delta b};\Rightarrow \|\boldsymbol{\delta x}\| \leq \left\|\boldsymbol{A}^{-1}\right\| \cdot \|\boldsymbol{\delta b}\| \tag{1.17}$$

and furthermore

$$\|\boldsymbol{b}\| \leq \|\boldsymbol{A}\| \cdot \boldsymbol{x};\Rightarrow \|\boldsymbol{x}\| \geq \frac{\|\boldsymbol{b}\|}{\|\boldsymbol{A}\|} \tag{1.18}$$

Therefore, from the above equations it follows that an upper bound for the relative error in **x** is given by (Moler, 2008)

$$\frac{\|\boldsymbol{\delta x}\|}{\|\boldsymbol{x}\|} \leq \|\boldsymbol{A}\| \left\|\boldsymbol{A}^{-1}\right\| \frac{\|\boldsymbol{\delta b}\|}{\|\boldsymbol{b}\|} \equiv \kappa(\boldsymbol{A}) \cdot \frac{\|\boldsymbol{\delta b}\|}{\|\boldsymbol{b}\|} \tag{1.19}$$

where $\kappa(\mathbf{A})$ is the condition number of matrix **A**. If $\kappa(\mathbf{A}) > > 1$, matrix **A** is said to be ill-conditioned. For example, if $\kappa(\mathbf{A})$ is 10^8, for instance, a relative error of 10^{-8} in the right-hand side could produce a relative solution error as large as 100%.

An important point to emphasize here is that the ill-conditioning characteristic of a matrix often originates from the very user who generates it, inadvertently. A simple way to produce ill-conditioned matrices is to have equations with coefficients spanning different orders of magnitude between one Eq. and another. There are no numerical procedures that guarantee a matrix will not present this problem.

The following options are available in Python to handle ill-conditioned matrix **A**:

(a) Use a least-squares solution of the system: **x**, **residuals**, rank, s = np.linalg.lstsq(**A**, **b**, rcond = None).
(b) Use the pseudo inverse solution: **x** = np.linalg.pinv(**A**) @ **b**. It is numerically stable and produces a solution of minimum norm. It is also useful when the system is overdetermined (more unknowns than equations), like Problem 5 in Sect. 1.5 below.

(c) For large dimension systems, apply an iterative routine, since this approach does not rely on the matrix inversion procedure, which is typically the source of ill-conditioning. Recommended options are lsqr, lsmr and cg within the scipy.sparse.linalg package.

For practical examples, see Sect. 1.5 below.

1.4 Sparse Matrices

Sparse matrices are defined as those in which the number of nonzero elements (nz) is much smaller than the total number of elements in the matrix (nz $<< n^2$ for square matrices). We have already seen some cases of sparse matrices with regular structure, such as the tridiagonal matrix presented in Sect. 1.2.2. The regular structure of the nonzero elements of that matrix allows solving the system of equations efficiently, reducing the number of operations and the storage space for the components of the sparse matrix, compared with solving a system of equations with a full matrix $\mathbf{A}$ containing nonzero elements. This type of regular structures in sparse matrices occurs in practice in two types of problems: (a) in stagewise separation processes (liquid–liquid extraction, distillation) where the mass balance in stage i only includes the compositions and flows associated with consecutive stages $i - 1$ to $i + 1$, and (b) in numerical methods where spatial derivatives are approximated by finite differences, which can be expressed in terms of the values of a function at adjacent spatial points. In both cases, it is possible to consider the regular structure of these matrices in the selection of an efficient solution method (Scott and Tuma, 2023).

In addition, sparse matrices also arise in the mass and energy balances of complex chemical processes that consist of numerous units, where these units are not simply connected to one another. In such cases, the resulting matrices are sparse, but they do not possess a simple structure. For examples of this type of matrices, it is recommended to visit the University of Florida sparse matrix repository (Davis & Hu, 2011; Kolodziej et al., 2019). The efficient solution of the linear equations associated with this type of sparse matrices requires the use of graph-based techniques, such as reordering of the equations and LU factorization algorithms that prevent the growth of nonzero elements. This approach is beyond the scope of this introductory book, but in the Sect. 1.5 some examples are provided to illustrate those specific applications. In Python, the package SciPy, specifically its sparse module (scipy.sparse), is an appropriate package for working with sparse matrices.

1.5 Problems

(1) Consider the matrices:

$$A = \begin{bmatrix} 1\,0 & -1 & 0 \\ 2\,5 & 1 & 0 \\ 1\,0 & 3 & 4 \\ 6\,0 & 0 & 6 \end{bmatrix} B = \begin{bmatrix} 1 & 0 \\ 2 & -5 \\ -1 & 0 \\ 4 & 6 \end{bmatrix}$$

Enter the matrices into the Python environment and indicate the results of applying the following operations:

(i) **x** = **A**[2,:]
(ii) **y** = **B**[:, 0]
(iii) **C** = **A**[1:4, 0:2]
(iv) **D** = np.diag(**A**)
(v) **E** = **B.T**
(vi) **C** = np.diag(np.diag(**A**))
(vii) **A**[2,:] @ **B**[:, 1]
(viii) **A**[1, 2] * **B**[3, 0]
(ix) *eigenvalues* = *np.linalg.eigvals*(**A**)
(x) **R**, **U** = np.linalg.eig(**A**)
(xi) **A** + 4 * np.ones((4, 4)) - np.eye(4)

(2) Consider the following matrix and vector

$$A = \begin{bmatrix} 10 & -2 & 3 & -1 \\ 5 & 11 & -3 & 8 \\ 2 & -3 & 15 & -2 \\ 4 & -4 & 0 & 9 \end{bmatrix} B = \begin{bmatrix} 1 \\ 3 \\ -3 \\ -1 \end{bmatrix}$$

Determine what would be obtained by performing the following operations in Python:

(i) **A** @ **B**
(ii) **C** = *np.diag(np.diag(***A**))
(iii) **A**[0:2, 2:4] @ **B**[2:4]
(iv) **A**[2,:] @ **B**
(v) **A**[:, 1].**T** @ **B**
(vi) **A**[1,2] * **B**[0]
(vii) **B.T** @ np.diag(**A**)
(viii) **R**, **U** = np.linalg.eig(**A** @ **A.T**)
(ix) **A**-5**ones*(4,4) + 3**eye*(4)

(3) For the matrices "**A**" from the two previous problems, decompose them so that they can be expressed in the following form: **A** = **D** + **L** + **U**, where **D** is the main diagonal of **A**, **L** is a lower triangular matrix, and **U** is an upper triangular matrix.

(4) Which of the following matrices are singular?

$$\boldsymbol{A}=\begin{bmatrix}1&0&-2&3\\3&1&1&4\\-1&0&2&-1\\4&2&6&0\end{bmatrix};\boldsymbol{B}=\begin{bmatrix}3&2&-1\\0&-1&4\\6&3&2\end{bmatrix};\boldsymbol{C}=\begin{bmatrix}1&0&-2&3\\3&1&1&4\\-1&0&2&-1\\4&3&6&0\end{bmatrix}$$

If the matrix is not singular, find its inverse.

(5) Find the slope and y-intercept of the least-squares line that passes through the points: (1, 1), (2, 1.5), and (3, 2.2). Plot your results.

(6) Do the following systems of equations have a solution? Find the solution if it exists.

$$\text{(a)}\quad\begin{bmatrix}2&1&3\\-1&0&2\\6&2&2\end{bmatrix}\cdot\boldsymbol{x}=\begin{bmatrix}1\\0\\0\end{bmatrix};\quad\text{(b)}\begin{bmatrix}1&1&0\\0&1&1\\1&0&1\\1&1&1\end{bmatrix}\cdot\boldsymbol{x}=\begin{bmatrix}1\\-2\\0\\4\end{bmatrix};$$

$$\text{(c)}\begin{bmatrix}1&3&1&-1\\2&0&1&1\\0&-1&4&1\\0&1&1&-5\end{bmatrix}\cdot\boldsymbol{x}=\begin{bmatrix}3\\1\\6\\16\end{bmatrix}$$

(7) Solve the following linear systems using Gauss-Jordan elimination with different pivoting options.

$$\text{(a)}\quad\begin{bmatrix}2x_1+x_2-3x_3\\-x_1+3x_2+2x_3\\3x_1+x_2-3x_3\end{bmatrix}=\begin{bmatrix}-1\\12\\0\end{bmatrix}\quad\text{(b)}\quad\begin{bmatrix}8x_1+x_2+x_3\\-x_1+7x_2+2x_3\\x_1\quad x_2+10x_3\end{bmatrix}=\begin{bmatrix}9\\7\\4\end{bmatrix}$$

(8) Use the Jacobi and Gauss–Seidel methods to solve the system of equations from the previous problem, case (b); in both cases use the initial guess $\mathbf{x}^{(0)}=[1,1,0.4]^{\mathrm{T}}$.

(9) For the system of linear equations:

$$\begin{bmatrix}1&1&1&1&1\\1&2&3&4&5\\1&3&6&10&15\\1&4&10&20&35\\1&5&15&35&70\end{bmatrix}\cdot x=\begin{bmatrix}0.1\\-0.2\\0\\0.2\\-0.1\end{bmatrix}$$

(a) Find the LU factorization of matrix **A**. Is this factorization suitable for numerically solving the system of equations? Explain.
(b) Use the results from (a) to compute the determinant of matrix **A** (note: the Python standard function cannot be used).
(c) Determine the solution **x** by explicitly using the results from (a).

(10) Construct a function that solves the linear system **A**@**x** = **b** by applying the Cholesky factorization **A** = $\mathbf{R}^T$@**R**, where **R** is an upper triangular matrix, using Python's *cholesky* function.

Requirements:

- The function should return the solution **x** if **A** is symmetric positive definite (SPD).
- If **A** is not SPD (i.e., has at least one nonpositive eigenvalue), the function should raise an error message. You can detect this by checking eigenvalues with numpy.linalg.eigvals(**A**).
- For a SPD **A**, verify that the solution is correct (e.g., by checking residual norm ||**A**@**x** − **b**|| is small).

  Test your function with the following linear systems. Verify that your function finds the correct solution.

$$A_1 = \begin{bmatrix} 5 & -2 & 4 \\ -2 & -3 & -1 \\ 4 & -1 & 6 \end{bmatrix}; \quad A_2 = \begin{bmatrix} 26 & -7 & 4 \\ -7 & 2 & -1 \\ 4 & -1 & 9 \end{bmatrix}; \quad A_3 = \begin{bmatrix} 45 & -8 & 46 \\ -8 & 14 & -11 \\ 46 & -11 & 53 \end{bmatrix};$$

$$b_1 = \begin{bmatrix} 1 \\ 0 \\ 5 \end{bmatrix}; \quad b_2 = \begin{bmatrix} 1 \\ 2 \\ -2 \end{bmatrix}; \quad b_3 = \begin{bmatrix} -3 \\ 1 \\ 4 \end{bmatrix}$$

(11) Tridiagonal linear systems arise often in Chemical Engineering Process modeling. Make a function that solves such a system using Gaussian elimination, taking advantage of the matrix structure to minimize the number of arithmetic operations. Test your routine by comparison with the standard solution provided by Python, for the following cases:

(i) **e** = np.ones((10, 1)); **data** = [**e,** − 2.5 * **e, e**]; diags = [−1, 0, 1]; **A** = spdiags([data, diags, 10, 10, format = 'csc'); **b** = $[-1\ 0\ 0 \ldots 0\ 0\ -1]^T$;
(ii) **e** = np.ones((20, 1)); **data** = [**e,** -2.5 * **e, e**]; diags = [−1, 0, 1]; **A** = spdiags([data, diags, 20, 20, format = 'csc'); **b** = $[0\ 0\ 0 \ldots 0\ 0\ \text{-}1]^T$;

(12) Consider the linear system **A**@**x** = **b** with

$$A = \begin{bmatrix} 6 & 6 & 3.00001 \\ 10 & 8 & 4.00003 \\ 6 & 4 & 2.00002 \end{bmatrix}; \quad b = \begin{bmatrix} 30 \\ 42 \\ 22 \end{bmatrix}$$

(a) Solve that system in Python using: x = np.linalg.solve(**A**, **b**). ¿Do you get the same result when applying: **x** = np.linalg.inv(**A**)@ **b**?
(b) Define a new right-hand side vector: $\mathbf{b}_1$ = [30.00002; 42.00006; 22.00004]; compute the new solution $\mathbf{x}_1$ using np.linalg.solve and np.linalg.inv. Are there significant differences between $\mathbf{x}_1$ and **x**? Explain.
(c) Compare the relative errors in the solution caused by the change in the right-hand side vector: norm($\mathbf{b}_1-\mathbf{b}$)/norm(**b**) and in the solution: norm($\mathbf{x}_1-\mathbf{x}$)/norm(**x**). What is going on here?
(d) Now apply the recommended options listed in Sect. 1.3 for handling ill-conditioned matrices. Compute **x** and $\mathbf{x}_1$ for the cases when the right-hand side vector is **b** and $\mathbf{b}_1$. What results do you get? How do these results compare with parts (a) and (b) above?

(13) Consider the matrix **A** = magic(5) and the vector **b** = np.sum(**A**, axis = 1). The system of equations **A**@**x** = **b** then has as its exact solution **x** = np.ones(5,1). Notice that **A** is ill conditioned. Test whether it is possible to numerically find the solution to this system of equations by trying the routines listed at the end of Sect. 1.3.

The following Python script generates the gallery matrices and prints magic(5).

```
import numpy as np

def magic(n):
    """
    Generates a magic square of order n (n must be odd)
    """
    A = np.zeros((n, n), dtype=int)

    i, j = 0, n // 2
    for num in range(1, n*n + 1):
        A[i, j] = num
        new_i = (i - 1) % n
        new_j = (j + 1) % n
        if A[new_i, new_j]:
            i = (i + 1) % n
        else:
            i, j = new_i, new_j

    return A

A = magic(5)
print(A)
```

(14) The following tridiagonal system of linear equations represents the equations of a simple model of a liquid–liquid extraction process, where the $x_{(i)}$ composition of a solute of interest changes throughout the n extraction steps (e.g., copper sulfate is extracted from a copper-poor aqueous solution and concentrated in an organic solvent stream).

$$
\begin{aligned}
&-1.6x_1 + 0.6x_2 = -0.04 \\
&x_{i-1} - 1.6x_i + 0.6x_{i+1} = 0;\ i = 2, 3, \ldots, n-1 \\
&x_{n-1} - 1.6x_n = -0.9
\end{aligned}
$$

Take $n = 15$ steps and calculate the vector of concentrations **x**. Show the results using plot(**x**). Use the Thomas algorithm code presented in Example 1.2.

(15) Consider the system of linear equations:

$$\begin{bmatrix} 7 & -1 & -2 & -3 \\ -1 & 7 & -3 & -2 \\ -2 & -3 & 7 & -1 \\ -3 & -2 & -1 & 7 \end{bmatrix} \begin{bmatrix} x_1 \\ x_2 \\ x_3 \\ x_4 \end{bmatrix} = \begin{bmatrix} 1 \\ 1 \\ 1 \\ 1 \end{bmatrix}$$

Whose exact solution is $\mathbf{x}^* = [1\ 1\ 1\ 1]^T$. Although matrix **A** has eigenvalues greater than 1.0, it is possible to propose the following iterative scheme of conjugate gradient (Sewell, 1988):

(i) Assume a random initial solution $\mathbf{x} = \text{rand}(n,1)$
(ii) Calculate the initial residue $\mathbf{r} = \mathbf{b}\text{-}\mathbf{A}@\mathbf{x}$, where in this case $b = \text{ones}(n,1)$
(iii) Calculate the initial vector $\mathbf{p} = \mathbf{r}$
(iv) Iteration cycle: Calculate successively:
 (1) lambda $= (\mathbf{r}^T @\mathbf{p})/(\mathbf{p}^T @ (\mathbf{A}@\mathbf{p}))$
 (2) $\mathbf{x} = \mathbf{x} +$ lambda*$\mathbf{p}$
 (3) $r = \mathbf{r} -$ lambda*($\mathbf{A}@\mathbf{p}$); if norm($\mathbf{r}$) == 0, stop because the solution has been found
 (4) alpha $= -(\mathbf{r}^T\ @\ (\mathbf{A}@\mathbf{p}))/(\mathbf{p}^T\ @\ (\mathbf{A}@\mathbf{p}))$
 (5) $p = r +$ alpha*$\mathbf{p}$
 (6) Go back to (1) and recalculate lambda, etc.

Iterate on this method, starting with $\mathbf{x}^{(0)} = \text{np.random.rand}(n,1)$, for which you must construct a function that receives matrix **A** and vector **b** as arguments and produces the iterated solution **x**. Convergence is reached when the mixed error is less than 10^{-6}.

(16) In applications of structural engineering, chemical process modeling and fluid mechanics, large linear systems ($N >> 1$) and with sparse matrices are usually generated. To solve these systems, Python has developed specialized routines for these types of problems.

(a) The *scipy.sparse.linalg.gmres* function solves a linear system by applying the GMRES algorithm (Saad and Schultz, 1986). Apply this routine to the case of the DK01R matrix (SuiteSparse Matrix Collection) that comes from a computational fluid dynamics problem, using a random vector as the right side. For this you will first have to generate a pre-conditioning of the DK01R array, using incomplete LU factorization. Get the number of iterations required to get convergence (with $\mathbf{x}^{(0)} = \text{np.zeros}(n,1)$, tol $= 1.0 \times 10^{-6}$).
(b) Compare the performance of the *scipy.sparse.linalg.bicgstab* function (Saad, 2003) to solve case (a), again using pre-conditioning with incomplete LU factorization, $\mathbf{x}^{(0)} = \text{np.zeros}(n,1)$, tol $= 1.0 \times 10^{-6}$.

(c) Compare both functions in terms of execution time, number of iterations performed and magnitude of final residual.
(d) The *scipy.sparse.linalg.bicg* function performs a bi-conjugate gradient iteration (Barret et al., 1994). Test this method with the non-symmetric matrix TOLS1090 (computational fluid dynamics, available in SuiteSparse Matrix Collection), using pre-conditioning with incomplete LU factorization and **b** is a random vector, $\mathbf{x}^{(0)}$ = np.zeros, tol = 1.0×10^{-8}.
(e) Repeat the case (d) using the *scipy.sparse.linalg.qmr* routine (Freund & Natchigal, 1991), using incomplete LU factorization for pre-conditioning.
(f) Compare both routines in terms of execution time, number of iterations performed, and magnitude of the final residual.

(17) When a matrix **A** is well conditioned, then the linear system of equations **A**@**x** = **b** can be transformed into the normal system of equations: $(\mathbf{A}^T@\mathbf{A})@\mathbf{x} = \mathbf{A}^T@\mathbf{b}$. In this case the conjugate gradient method can be applied to the normal system of equations, since the matrix $(\mathbf{A}^T@\mathbf{A})$ is symmetric.

(a) The *random.rand* function generates a matrix of random numbers of dimension $m \times n$, with a given density (fraction) of non-zero elements and setting the reciprocal of the condition number (rc) of the matrix **A** thus generated. Using a tolerance of 10^{-8}, a maximum number of iterations of 500, a right-hand side vector of random numbers and $\mathbf{x}^{(0)} = [1\ 1 \ldots.\ 1]^T$, calculate the final residual and the number of iterations performed by *scipy.sparse.linalg.cg* for the following matrices:
 (i) $N = M = 100$, density = 0.04, rc = 0.1
 (ii) $N = M = 300$, density = 0.08, rc = 0.01
 (iii) $N = M = 800$, density = 0.16, rc = 0.001
The following Python script generates the first of the above matrices

```
import numpy as np
from scipy.sparse import random

A = random(
    100, 100,
    density=0.03,
    data_rvs=lambda k: 0.1 * np.random.rand(k)
)
```

In all cases solve the normal system of equations and use pre-conditioning, using ILU incomplete factorization: spilu(**A**, drop_tol = 0.0, fill_factor = 1). Indicate the calculation time in each case (including the calculation time associated with running *ilu*).
(b) The command R = random(n, n, density = 0.1, data_rvs = np.random.rand) followed by: $R = (R + R\cdot\text{T})/2$ generates a sparse, symmetrical matrix of dimension $n \times n$, where density is the fraction of non-zero elements (0.1 in this example). The values generated are normally distributed with mean 0 and variance 1. Using random, generate

a matrix of dimension 200×200, with 2% non-zeros. Solve with the conjugate gradient method (*scipy.sparse.linalg.cg*). Use a random vector **b**, random $\mathbf{x}^{(0)}$, tol $= 1.0 \times 10^{-9}$, and maxit $= 900$. Compare execution times for cases without and with pre-conditioning. Using Incomplete LU Factorization to find pre-conditioning matrices.

(18) The matrix **A** and vector **b** contained in the data structure "**nos4.mat**" is a sparse matrix of dimension 100×100 that comes from the numerical analysis of a solid structure (https://sparse.tamu.edu/). It is proposed to find a solution to the problem A*$\mathbf{x} = \mathbf{b}$ by means of an iterative method with the *scipy.sparse.linalg.cg* routine. Using tolerance $= 10^{-8}$, solve the system without and with pre-conditioning. Do both methods converge?

(19) The matrix **A** and vector **b** that are included in the data structure "**cavity04.mat**" correspond to the analysis of the flow in a cavity at low Reynolds number (Re = 600), which leads to a system of linear equations (https://sparse.tamu.edu/). Matrix A has a dimension of 317×317 and its condition number is greater than 10^7. Then, it is proposed that you test whether the *scipy.sparse.linalg.gmres* routine can solve the system of equations with a tolerance $= 10^{-8}$ and using pre-conditioning with incomplete LU factorization. If the routine converges, graph the solution vector and indicate how many iterations were made.

References

Axelson, O. (1994). *Iterative solution methods*. New York: Cambridge University Press.

Barrett, R., Berry, M., & Chan, T. F. (1994). *Templates for the solution of linear systems: Building blocks for iterative methods*. Philadelphia: SIAM.

Davis, T. A., & Hu, Y. (2011). The University of Florida sparse matrix collection. *ACM Transactions on Mathematical Software, 38*(1), 2026. https://doi.org/10.1145/2049662.2049663(Accessed: 10January

Freund, R. W., & Nachtigal, N. M. (1991). QMR: A quasi-minimal residual method for non-Hermitian linear systems. *SIAM Journal Numerische Mathematik, 60*, 315–339.

Gerald, C. F. (1998). *Applied numerical analysis* (6th ed.). Addison-Wesley.

Hanna, O. W., & Sandall, O. C. (1995). *Computational methods in chemical engineering*. New Jersey: Prentice Hall.

Jorquera, H., & Gelmi, C. (2014). *Métodos Numéricos Aplicados a Ingeniería. Casos de estudio usando MATLAB®*. Ediciones UC.

Kolodziej, S. P. et al. (2019). The SuiteSparse matrix collection website interface. *Journal of Open Source Software, 4*(35), 1244–1248. https://doi.org/10.21105/joss.01244 (Accessed: 10 January 2026).

Moler, C. (2008). *Numerical Computing with MATLAB*, 1st Ed. http://www.mathworks.com/moler/. Accessed 10 January 2026.

Nakamura, S. (1991). *Applied numerical methods with software*. Prentice Hall.

Saad, Y., & Schultz, M. H. (1986). GMRES: A generalized minimal residual algorithm for solving nonsymmetric linear systems. *SIAM Journal on Scientific and Statistical Computing, 7*(3), 856–869.

Saad, Y. (2003). *Iterative Methods for Sparse Linear Systems, Second edition*. SIAM.
Scott, J., & Tuma, M. (2023) *Algorithms for Sparse Linear Systems*. Springer Nature.
Sewell, G. (1988). *The numerical solution of ordinary and partial differential equations*. Academic Press.
Wang, J. C., & Henke, G. E. (1966). Tridiagonal matrix for distillation. *Hydrocarbon Process, 45*(8), 155–163.

Chapter 2
Nonlinear Equations

To solve algebraic equations in general, iterative solution methods are employed. An iterative method consists of the following stages (see Fig. 2.1):

(i) Estimating an initial value for the sought solution.
(ii) A formula for updating the obtained approximate solution.
(iii) A criterion to terminate the updating process (convergence check).

Notes:

(a) The initial stage often requires including tests to validate the consistency of the initial data (e.g., positive concentrations).
(b) It is important to distinguish between the complete iterative process (or algorithm) and the iteration formula (or update formula).
(c) The verification of convergence for the iterative process is essential and must anticipate all possible outputs of the iterative method. It should be able to detect when the algorithm fails. The customary approach consists in selecting a type of error to be computed, which may be the absolute or relative error of the current solution estimate, as compared to the previous estimate. Absolute error makes sense when a solution is much larger than 1.0 (example: an absolute temperature), while relative error is preferred when a solution is much smaller than 1.0 (example: molar fraction at equilibrium).

We now present the most common methods for solving nonlinear equations in a single variable.

2.1 Fixed-Point Method

This method uses only the most recently estimated value to obtain the next approximation of the solution. For this reason, it is sometimes referred to as functional iteration. Consider the solution of the scalar equation

H. Jorquera and C. A. Gelmi, *Numerical Methods in Chemical Process Engineering Using Python*, https://doi.org/10.1007/978-3-032-22958-8_2

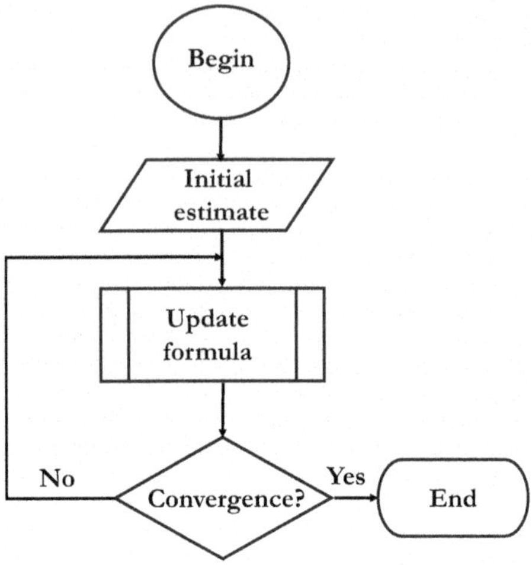

Fig. 2.1 Typical flow diagram of an iteration process

$$f(x) = 0. \tag{2.1}$$

We now compute a sequence $\{x^{(k)}\}\ k = 0, 1, 2, \ldots$, which, if successful, converges to the solution x^*. The user must provide the initial estimate $x^{(0)}$, which is not a trivial step. The subsequent approximations are given by

$$x^{(k+1)} = \phi\left(x^{(k)}\right); \quad k = 0, 1, \ldots \tag{2.2}$$

where $x = \phi(x)$ shares the same solution as $f(x) = 0$, that is, $x^* = \phi(x^*)$.

The iteration (or update) formula $x = \phi(x)$ is obtained by rearranging terms or by adding terms to both sides of the original equation $f(x) = 0$. This process of moving from $f(x) = 0$ to $x = \phi(x)$ is not unique, as seen in the following example.

Example 2.1 Fixed-point method

Consider the equation: $f(x) = x^3 - 7x - 6 = 0$. Some possible iteration functions are:

$$\phi_1(x) = (x^3 - 6)/7$$
$$\phi_2(x) = (7x + 6)/x^2$$
$$\phi_3(x) = (2x^3 + 6)/(3x^2 - 7).$$

As we know a priori that $f(x) = (x + 1)(x + 2)(x - 3)$, what we do is test the performance of the iteration formulas by choosing the initial value $x^{(0)} = -1.1, -2.2$ and 4.0, respectively.

Carrying out the calculations with the different iteration functions, for a maximum absolute error of 10^{-5} in the solution, the results are shown in Table 2.1.

Table 2.1 Comparison of results x^* (iterations) for different update functions

Initial estimate $x^{(0)}$	$\phi_1(x)$	$\phi_2(x)$	$\phi_3(x)$
− 1.1	− 1 (12)	− 2 (10)	−1 (4)
− 2.2	Diverges	− 2 (9)	− 2 (4)
4.0	Diverges	Diverges	3 (5)

The following is the Python code used to generate the fixed-point method and its application to the first case with $\phi_1(x)$. The reader is encouraged to complete the code to verify all entries in Table 2.1

```
# Application to first fixed point function in example 2.1
# Define the fixed-point iteration
def fixed_point(g, x0, tol=1e-6, maxit=1000):
    x = x0
    for nit in range(1, maxit + 1):
        x_next = g(x)
        err = abs(x_next - x)
        if err < tol:
            return x_next, err, nit
        x = x_next
    # If not converged within max iterations, return last values
    return x, err, maxit

    # Define the fixed-point function
f1 = lambda x: (x**3 - 6) / 7

tol = 1e-5
maxit = 100
x0 = -1.1

root, err, nit = fixed_point(f1, x0, tol, maxit)

print('root = ', root)
print('err = ', err)
print('iter = ', nit)
```

Definition A function ϕ: $R \longrightarrow R$ is contraction in a set $D \in R$ if there exists a constant $0 < L < 1$ such that:

$$|\phi(a) - \phi(b)| < L \cdot |a - b|, \quad \forall a, b \in D \tag{2.3}$$

Graphically, this means that a given interval of the domain D is transformed, by the application of ϕ, into a subset of D. Let us view this in the following example.

Example 2.2 Convergence of the fixed-point method

Let us again consider $\phi_1(x) = (x^3{-}6)/7$ about the origin $x = 0$. Graphically, we can observe that the interval [0, 2] is mapped onto the interval [−6/7, 2/7], which clearly is not a subset of [0, 2], hence ϕ_1 is not a contraction there. However, it is also possible to observe graphically that

$$x \in [-2,\ 0] \Rightarrow \phi_1(x) \in [-2,\ -6/7], \therefore \phi_1(x) \text{ is a contraction in } D \equiv [-2,\ 0]$$

Below we present the Python script used to generate Figure 2.2.

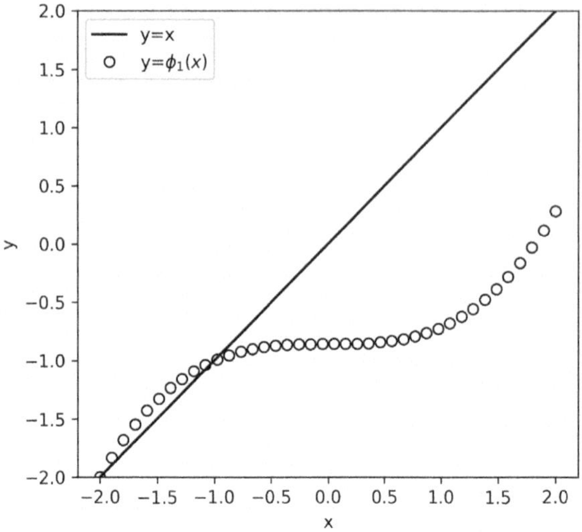

Fig. 2.2 Update function $\phi_1(x)$ for Example 2.2

```
import numpy as np
import matplotlib.pyplot as plt

# Figure 2.2
f1 = lambda x: x
f2 = lambda x: (x**3 - 6) / 7

x = np.linspace(-2, 2, 800)

plt.figure()
plt.plot(x, f1(x), '-k', label='y=x')
plt.plot(x, f2(x), 'k.', label=r'y=$\phi_1(x)$')
plt.ylim([-2, 2])
plt.gca().set_aspect('equal', adjustable='box')
plt.xlabel('x')
plt.ylabel('y')
plt.legend()

# Save figure
filename = 'Figure_2_2.png'
plt.savefig(filename, dpi=300, bbox_inches='tight')
plt.show()
```

Note that the contraction concept is intimately linked to the domain of function ϕ_1, and this domain is assumed to contain the solution sought a priori, so our physical intuition can provide an estimate of what this interval should be, to apply the definition.

Next, we establish the fundamental mathematical result on which this iterative method is based.

2.2 The Contraction Function Theorem (or Fixed-Point Theorem)

Let ϕ be a continuous function ϕ: $R \to R$ and let $D \subset R$ be such that: $x \in D \Rightarrow \phi(x) \in D$.

Suppose moreover that there exists a constant L with $0 < L < 1$ such that

$$|\phi(a) - \phi(\boldsymbol{b})| < \boldsymbol{L} \cdot |\boldsymbol{a} - \boldsymbol{b}|, \quad \forall \boldsymbol{a}, \boldsymbol{b} \in \boldsymbol{D} \tag{2.4}$$

Then, for every $x^{(1)} \in D$, the sequence Eq. (2.2) converges to a unique fixed-point $x^* \in D$.

Graphically, the iteration sequence Eq. (2.2) drives the distance between an estimate $x^{(k)}$ and the solution x^* to become progressively smaller, until it tends to zero, because the Lipschitz constant L is less than 1.0. However, for convergence to occur, all conditions of the foregoing theorem must be satisfied, which can be easily verified using the derivative of $\phi(x)$ as shown in the following example.

Example 2.3 Verification of convergence of the fixed-point method

Consider $f(x) = x^3 - 7x - 6 = 0$ and $\phi_1(x) = (x^3 - 6)/7$ to locate the root $x^* = -1$.

(a) from the previous example, we know that ϕ_1 is a contraction on $D = [-2, 0]$ and $x^* \in D$ via the mean value theorem it holds:

$$\phi_1(a) - \phi_1(b) = \phi_1'(\xi)(a - b); \quad a \le \xi \le b$$

so:

$$|\phi_1(a) - \phi_1(b)| \le \max_{\{\xi \in \Delta\}} |\phi_1'(\xi)| |a - b|$$

Por lo que tenemos que $L = \max\{\xi \in \Delta\} |\phi_1'(\xi)|$

$$\phi_1'(\xi) = 3/7\xi^2 \quad \text{and} \quad |\phi_1'(\xi)| < 1 \to \xi \in [-\sqrt{7/3}, \sqrt{7/3}]$$

The contraction condition is satisfied when we consider the intersection of the two intervals; consequently, the convergence region is $[-\sqrt{7/3}, 0]$. Since this region contains $x^* = -1$, the fixed-point iteration with $\phi_1(x)$ will converge to $x^* = -1$ for any initial value $x(1) \in [-\sqrt{7/3}, 0]$.

Verify this using the fixed-point function provided in the Example 2.2. Note that $\phi_1(x)$ cannot compute the root $x^* = -2$. Why?

2.2.1 Graphical Representation of the Fixed-Point Iteration

For the case of a scalar equation, the behavior of the iteration $x^{(k+1)} = \phi(x^{(k)})$ can be represented as a trajectory that connects the graphs of $y = \phi(x)$ and the horizontal axis, as shown in Fig. 2.3. Note that the solution sought corresponds to the crossing of the identity with the update function $\phi(x)$. It is possible to observe that the magnitude and sign of the derivative $\phi'(x)$ determine the success or failure of the fixed-point iteration:

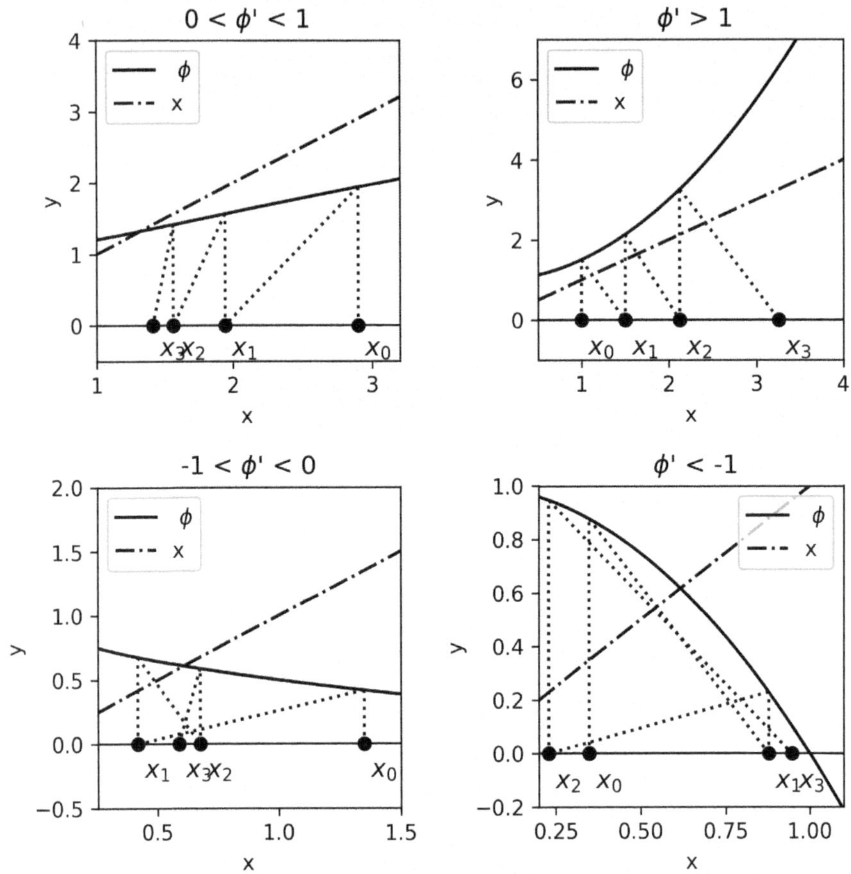

Fig. 2.3 Possible results for a fixed-point iteration

(a) In the upper panels in Fig. 2.3, if $\phi'(x) < 1$, then the sequence of estimates converges monotonously to the solution x^*. In contrast, if $\phi'(x) > 1$, the iteration quickly diverges.
(b) In the lower panels in Fig. 2.3, if $-1 < \phi'(x) < 0$, then the sequence of estimates converges to the solution x^* in an alternating direction pathway. For $\phi'(x) < -1$, the alternating path diverges quickly from the solution sought.

2.3 Interpolation Methods

The idea is that the function $f(x) = 0$ can be expanded around $x = x_1$ using a Taylor series; the series is truncated, and the truncated approximation to $f(x)$ is then used to locate its zero(s).

$$f(x) = f(x_1) + (x - x_1) \cdot f'(x_1) + (x - x_1)^2 \cdot f''(x_1)/2! + \cdots \tag{2.5}$$

2.3.1 *Linear Interpolation (Newton's Method)*

We expand $f(x)$ around $x = x^{(k)}$; denoting $P(x)$ the linear approximation:

$$P(x) = f\left(x^{(k)}\right) + \left(x - x^{(k)}\right) \cdot f'\left(x^{(k)}\right) \approx f(x) \tag{2.6}$$

And the new approximation $x^{(k+1)}$ is the root of $P(x) = 0$:

$$x^{(k+1)} = x^{(k)} - \frac{f\left(x^{(k)}\right)}{f'\left(x^{(k)}\right)};\ k = 0, 1, 2, \ldots \tag{2.7}$$

Example 2.4 Visualization of Newton's method convergence

Newton's method corresponds to a sequence of linear interpolations, as illustrated in Fig. 2.4 for the case $f(x) = x - x^2 + 0.5x^3$, starting with $x^{(0)} = 3.5$. The sequence of estimates is shown as black dots, and it is a monotonous sequence for this example.

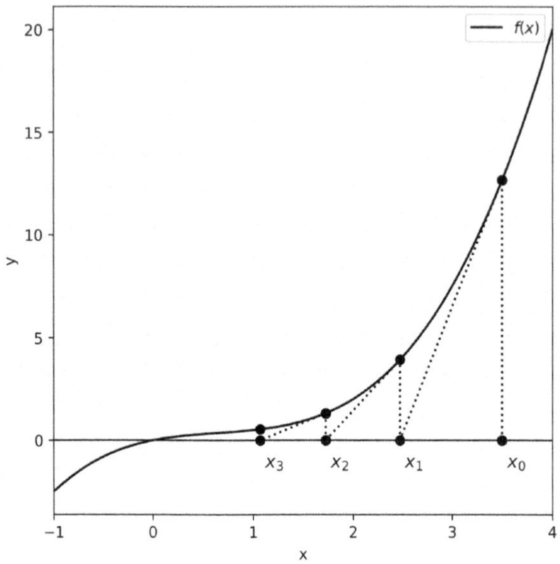

Fig. 2.4 Visualization of Newton's method

The following presents the Python script used to generate Fig. 2.4.

```
import numpy as np
import matplotlib.pyplot as plt

def f1(x):
    """Nonlinear function."""
    return x - x**2 + 0.5 * x**3

def f1p(x):
    """Derivative of f1."""
    return 1 - 2*x + 1.5 * x**2

# Function plot
plt.figure(figsize=(6, 6))

xx = np.linspace(-2, 4, 1000)
plt.plot(xx, f1(xx), "k-", label=r"$f(x)$")

plt.xlabel("x")
plt.ylabel("y")
plt.axhline(0, color="k", linewidth=1)
plt.xlim([-2, 4])

# Iterative sequence of Newton's method
x = np.zeros(4)
f = np.zeros(4)

x[0] = 3.5
f[0] = f1(x[0])
```

```
# Plot initial point
plt.plot(x[0], f[0], "ro")
plt.plot(x[0], 0, "ro")

# Label initial iteration: x_0
plt.text(x[0], 0, r"$\;x_0$", fontsize=11, verticalalignment="bottom")

for i in range(1, 4):
    # Newton iteration
    x[i] = x[i - 1] - f[i - 1] / f1p(x[i - 1])
    f[i] = f1(x[i])

    # Plot Newton construction lines
    plt.plot([x[i - 1], x[i - 1]], [0.0, f[i - 1]], "k:")
    plt.plot([x[i - 1], x[i]], [f[i - 1], 0.0], "k:")

    # Plot iteration points
    plt.plot(x[i], f[i], "ko")
    plt.plot(x[i], 0, "ko")

    # Label iteration: x_i
    plt.text(x[i], 0, rf"$\;x_{i}$", fontsize=11, verticalalignment="bottom")

plt.legend()
plt.show()
```

Notes:

(i) The update function is: $\phi(x) = x - f(x)/f'(x)$.
(ii) The method is of second order: quadratically convergent; this means that the number of correct decimal digits doubles at each iteration as $x(k) \to x^*$.

We can observe in Eq. 2.7, which defines the Newton update, that problems can arise near the solution if the derivative f' is zero or very small. The following theorem clarifies what can be expected from the iteration 2.7.

Theorem If the root x^* of $f(x) = 0$ lies in an interval I where $f'(x) \neq 0$, and furthermore f' is continuous on I and the convexity condition is satisfied:

$$f(a) - f(b) \geq f'(b) \cdot (a - b) \quad \forall a, b \in I \tag{2.8}$$

Then, for every $x^{(0)} \in I$, the Newton iteration produces a sequence $\{x^{(k)}\}$ that satisfies exactly one of the following conditions:

(i) $x^{(k)} \in I$ and the sequence $\{x^{(k)}\}$ converges monotonically to x^*.
(ii) $f(x^{(0)}) \cdot f(x^{(1)}) < 0$ and the sequence $\{x^{(k)}\}$ converges monotonically to x^*.
(iii) $x^{(1)} \notin I$ (if $I = R$, this option is not valid).

Corollary 1 The convexity condition can be replaced by a concavity condition:

$$f(a) - f(b) \leq f'(b) \cdot (a - b) \quad \forall a, b \in I \tag{2.9}$$

And the theorem remains valid.

Corollary 2 Instead of convexity or concavity conditions, the following condition can be used: $f''(x) \neq 0$ and is continuous for all $x \in I$.

2.3.2 Quadratic Interpolation

Given the issues of the classical Newton method when the derivative changes sign and the iteration becomes unstable, schemes have been proposed to avoid division by the derivative $f'(x)$.

In the quadratic interpolation approach, $f(x)$ is approximated around $x = x^{(k)}$ by a second-order polynomial $P(x)$:

$$P(x) = f\left(x^{(k)}\right) + \left(x - x^{(k)}\right) \cdot f'\left(x^{(k)}\right) + (x - x^{(k)})^2 \cdot f''\left(x^{(k)}\right)/2 \tag{2.10}$$

And $x^{(k+1)}$ is obtained by solving $P(x^{(k+1)}) = 0$:

$$x^{(k+1)} = x^{(k)} + \frac{-f'\left(x^{(k)}\right) \mp \sqrt{\gamma^{(k)}}}{f''\left(x^{(k)}\right)}; \gamma^{(k)} = \left[f'\left(x^{(k)}\right)\right]^2 - 2f\left(x^{(k)}\right)f''\left(x^{(k)}\right) \tag{2.11}$$

Since it is assumed that $x^{(k+1)} \approx x^{(k)}$, the root with the smallest absolute value is chosen in the iteration formula.

Notes:

(i) The cancelation of terms with similar absolute values but opposite signs can lead to a loss of significant digits in arithmetic.
(ii) Because it cannot be guaranteed in general that the discriminant $\gamma^{(k)}$ in Eq. (2.11) is positive, one of the factors $(x - x^{(k)})$ is replaced by the expression from the classical Newton method 2.7, thereby obtaining the extended Newton method:

$$x^{(k+1)} = x^{(k)} - \frac{f\left(x^{(k)}\right)}{f'\left(x^{(k)}\right) - 0.5f''\left(x^{(k)}\right)f\left(x^{(k)}\right)/f'\left(x^{(k)}\right)} \tag{2.12}$$

Example 2.5 shows that this method helps reduce the oscillations that occur with the classical Newton method, although one cannot conclude that Eq. (2.12) is always robust with respect to the initial estimate $x^{(0)}$.

Example 2.5 Comparison of the classical Newton method and its extended version

Compare the performance of the classical Newton method with its extended variant for the case $f(x) = x^3 - x - 1$, starting from $x^{(0)} = -1$.

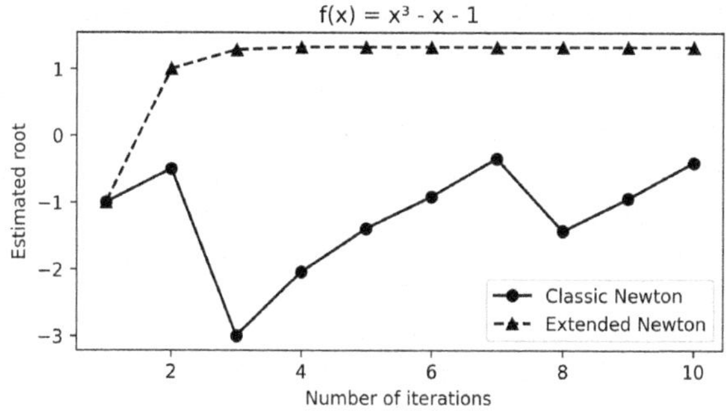

Fig. 2.5 Performance of the classical Newton iteration and the extended version

Figure 2.5 is obtained, showing that the classical method oscillates during the first 10 iterations without approaching the correct root, whereas the extended method requires only 4 or 5 iterations to reach the correct root, even though the derivative near $x^{(0)} = -1$ changes sign.

The following presents the Python code used to perform the calculations and produce the preceding figure.

```
import numpy as np
import matplotlib.pyplot as plt

# Script that solves Example 2.5
def f(x):
    return x**3 - x - 1

def fp(x):
    return 3*x**2 - 1

def fpp(x):
    return 6*x

n_iter = 10

# Classic Newton
xN = np.zeros(n_inter)
fN = np.zeros(n_iter)
xN[0] =  1.0
fN[0] = f(xN[0])

# Extended Newton (quadratic variant)
xC = np.zeros(n_iter)
fC = np.zeros(n_inter)
xC[0] = -1.0
fC[0] = f(xC[0])

for i in range(1, n_iter):
    # Classic Newton iteration
    xN[i] = xN[i - 1] - fN[i - 1] / fp(xN[i - 1])
    fN[i] = f(xN[i])
```

```
        # Extended Newton iteration
        der = fp(xC[i - 1])
        xC[i] = xC[i - 1] - fC[i - 1] / (
            der - 0.5 * fC[i - 1] * fpp(xC[i - 1]) / der
        )
        fC[i] = f(xC[i])

# Comparison plot
iters = np.arange(1, n_iter + 1)

plt.figure()
plt.plot(iters, xN, "k-o", label="Classic Newton")
plt.plot(iters, xC, "k--^", label="Extended Newton")
plt.gca().set_aspect("equal", adjustable="box")
plt.title("f(x) = x³ - x - 1")
plt.xlabel("Number of iterations")
plt.ylabel("Estimated root")
plt.legend()

# Save figure
filename = 'Figure_2_5.png'
plt.savefig(filename, dpi=300, bbox_inches='tight')
plt.show()
```

2.3.3 Python Implemented Routines for Scalar Equations

In Python, a general-purpose routine available is scipy.optimize.root_scalar. This routine allows users to choose among several techniques to achieve convergence.

For the case of polynomials, Python provides the *np.roots* function, which computes all roots. If one wishes to find a zero of an equation when there is real data with measurement errors, the original problem Eq. (2.1) can be posed as the minimization of f^2 and the scalar minimization function *minimize_scalar* from package *scipy.optimize* can then be applied.

2.4 Systems of Equations: Newton's Method and Its Variants

The vectorial nature of a system of equations, in contrast to the scalar character of single-variable equations discussed previously, does not substantially alter the way the problem is solved iteratively. Nevertheless, we will highlight some details that must be considered when implementing the various algorithms.

We have already seen that Newton's method corresponds to a local linear interpolation (because the Taylor expansion is local). In the same vein, we proceed for systems of equations: we approximate

$$\boldsymbol{f}(\boldsymbol{x}) \approx \boldsymbol{f}\left(\boldsymbol{x}^{(k)}\right) + \boldsymbol{J}\left(\boldsymbol{x}^{(k)}\right) \cdot \left(\boldsymbol{x} - \boldsymbol{x}^{(k)}\right) = \mathbf{0} \tag{2.13}$$

where the gradient, or Jacobian $\mathbf{J}$, of f is given by

Table 2.2 Typical Newton method algorithm

(i)	Choose initial vector $\mathbf{x}^{(0)}$
(ii)	Initialize iteration counter $k = 0$
(iii)	Checking for convergence; if it is fulfilled, stop
(iv)	Solve linear system $\mathbf{J}(\mathbf{x}^{(k)})\cdot\mathbf{d}^{(k)} = -\mathbf{f}^{(k)}$
(v)	Update $\mathbf{x}^{(k+1)} = \mathbf{x}^{(k)} + \mathbf{d}^{(k)}$
(vi)	Update error
(vii)	$k = k + 1$
(viii)	Back to (iii)

$$J_{ij} = (\nabla f)_{ij} = \frac{\partial f_i}{\partial x_j} \tag{2.14}$$

Therefore, to obtain the new value $x^{(k+1)}$ we must solve the linear system

$$\boldsymbol{J}\left(\boldsymbol{x}^{(k)}\right) \cdot \left(\boldsymbol{d}^{(k)}\right) = -\boldsymbol{f}\left(\boldsymbol{x}^{(k)}\right) \tag{2.15}$$

For the correction $\mathbf{d}^{(k)} = \mathbf{x}^{(k+1)} - \mathbf{x}^{(k)}$. This is Newton's method for N variables. Note that it is necessary for the Jacobian inverse to exist (i.e., no singularity). Table 2.2 presents the standard algorithm for the method in N variables.

2.4.1 Variations of the Newton Method

In Eq. 2.15, $\mathbf{J}^{(k)}$ must be evaluated at every iteration using expressions for the partial derivatives. For systems with many equations, this is not practical. Several alternatives exist to reduce the amount of work or to simplify the solution technique.

(a) We can replace $\mathbf{J}^{(k)}$ with a constant matrix $\mathbf{A}$, which yields the parallel-string method:

$$\boldsymbol{A} \cdot \boldsymbol{d}^{(k)} = -\boldsymbol{f}^{(k)} \tag{2.16}$$

(b) If we choose $\mathbf{A} = \mathbf{I}$, we have a fixed-point iteration on each variable separately, with loss of convergence speed. A more appropriate choice is $\mathbf{A} = \mathbf{J}^{(0)}$, so the iteration is

$$\boldsymbol{J}^{(0)} \cdot \boldsymbol{d}^{(k)} = -\boldsymbol{f}^{(k)}. \tag{2.17}$$

This is frequently used in solving ordinary differential equations by implicit methods (Chap. 3). Generally, the Jacobian $\mathbf{J}$ is re-evaluated every time it is to improve the speed of convergence.

(c) When partial derivatives are not explicitly given, the Jacobian can be determined by finite differences. This leads to the discrete Newton method:

$$\boldsymbol{\Delta}^{(0)} \cdot \boldsymbol{d}^{(k)} = -\boldsymbol{f}^{(k)}. \tag{2.18}$$

Comments:

From the approaches just discussed, two drawbacks emerge:

(i) The Jacobian must be evaluated (sometimes by finite differences). This entails n^2 operations per step (which can be reduced if the Jacobian is updated every m iterations).
(ii) The solution for $\mathbf{d}^{(k)}$ involves $n^3/3$ operations per iteration, although if the Jacobian is fixed, this amounts to n^2 operations/iteration.

In general, the function **f** is expensive to evaluate in process engineering problems, so Newton methods are costly in terms of computational time. To reduce the number of operations per iteration, quasi-Newton methods are employed, which use an approximation to the inverse of the Jacobian that is updated at each step to reduce the operation count. For more details on these methods, see the article by Sargent (1981) and the work of Dennis and Mori (1977).

Example 2.6 shows a Python code to solve a system of N equations using the discrete Newton method (2.18). Each column 'j' of the Jacobian is computed by introducing a perturbation to the j-th variable via finite differences.

Example 2.6 Application of the discrete Newton method in N variables

The goal is to solve the following system of nonlinear equations:

$$f_1(x) = x_1^2 + x_2^2 - x_3 - 2 = 0$$
$$f_2(x) = x_1 + 5x_2 + 1 = 0$$
$$f_3(x) = x_1x_3 - 2x_1 + 1 = 0$$

Using the discrete Newton method. The programmed method must ensure that the number of iterations does not exceed a maximum value and use as an error criterion the norm of the vector **f**: err = norm($\mathbf{f}^{(k)}$) which must be less than 10^{-5}. How many distinct roots do you find?

Table 2.3 shows the results of applying the discrete Newton method to the above system of equations, with different initial vectors $\mathbf{x}^{(1)}$. In this case, we find three different roots. This is because if we eliminate x_2 and x_3 in the system's equations, we find a cubic equation for x_1, which has three distinct real roots.

Table 2.3 Solution of Example 2.6

Initial estimate $\mathbf{x}^{(1)}$	Solution $\mathbf{x}$	norm($\mathbf{f}(\mathbf{x})$)	No. of iterations
$\begin{bmatrix} 0 & 0 & 0 \end{bmatrix}^T$	$[0.2584\ -0.2517\ -1.8699]^T$	3.40×10^{-11}	4
$[1\ 1\ 1]^T$	$[1.7686\ -0.5537\ 1.4346]^T$	9.77×10^{-15}	8
$[-1\ 4\ 5]^T$	$[-2.1039\ 0.2208\ 2.4753]^T$	1.63×10^{-13}	6

Below is the list of the Python script that solves the first case in Example 2.6.

```
import numpy as np

def newton_discrete_jacobian(fun, x1, tol=1e-8, maxit=50, ep=1e-6):
    xr = np.asarray(x1, dtype=float).reshape(-1)
    N = xr.size
    I = np.eye(N)

    for it in range(1, maxit + 1):
        xrold = xr.copy()
        f_old = np.asarray(fun(xrold), dtype=float).reshape(-1)

        # Finite-difference Jacobian
        J = np.zeros((N, N))
        for j in range(N):
            f_pert = np.asarray(fun(xrold + ep * I[:, j]), dtype=float).reshape(-1)
            J[:, j] = (f_pert - f_old) / ep

        # Newton step
        dx = np.linalg.solve(J, f_old)
        xr = xrold - dx

        f_new = np.asarray(fun(xr), dtype=float).reshape(-1)
        err = np.linalg.norm(f_new)

        print(f"Iteration {it:2d} | ||f|| = {err:.3e} | x = {xr}")

        if err < tol:
            break

    return xr, f_new, err, it

def f1(x):
    x = np.asarray(x).reshape(-1)
    y = np.empty(3)
    y[0] = x[0]**2 + x[1]**2 - x[2] - 2
    y[1] = x[0] + 5*x[1] + 1
    y[2] = x[0]*x[2] - 2*x[0] + 1
    return y

# Run the example
x0 = np.array([0.0, 0.0, 0.0])  # initial guess

root, fval, err, nit = newton_discrete_jacobian(f1, x0)

print("\nFinal result")
print("------------")
print("Estimated root:", root)
print("f(root):", fval)
print("||f||:", err)
print("Number of iterations:", nit)
```

2.4.2 Python Implemented Routines for Systems of Equations

In the case of systems of nonlinear equations, Python provides the basic solution routine *fsolve* from package *SciPy.Optimize*. This routine requires a Python function that expresses the system of equations as a function of the vector x, a second (required) argument is an initial estimate of the root. The Jacobian of the system of equations may be optionally used as another Python function, being a third input to *fsolve*, and this usually improves convergence. Alternatively, the function *root* can employ additional algorithms, like quasi-Newton methods, etc.

If this routine has difficulty finding a solution, especially as the dimension of x increases, one can attempt to minimize the scalar function $f(x)^T f(x)$, which attains a minimum when $f(x) = 0$. In this way, one may try *minimize*, which minimizes the scalar function of the vector x. This function has the flexibility to run without constraints, or to add constraints and thus reduces the search space for the solution.

2.5 Problems

(1) Consider the solution of the cubic equation: $f(x) = x^3 - x^2 - 2 = 0$.

It is proposed to find a positive root (>1) of the above equation, using the following update functions:

(a) Taking the cube root, call it $\phi_1(x) = (x^2 + 2)^{1/3}$
(b) Taking the square root, call it $\phi_2(x) = (x^3 - 2)^{1/2}$

Iterate on the fixed point using the two update functions described above. Indicate whether iterations are converging and why such behavior occurs. Is it possible to improve the performance of these two functions by using the Aitken acceleration method?

Note: Aitken's method (Aitken, 1926) evaluates the new estimate of the solution every two fixed-point estimates, i.e.:

$$x^{(1)} = \phi\left(x^{(0)}\right);\, x^{(2)} = \phi\left(x^{(1)}\right);\, x^{(3)} = \frac{\left(x^{(2)} - x^{(1)}\right)^2}{x^{(0)} - 2x^{(1)} + x^{(2)}};$$

$$x^{(4)} = \phi\left(x^{(3)}\right);\, x^{(5)} = \phi\left(x^{(4)}\right);\, x^{(6)} = \frac{\left(x^{(5)} - x^{(4)}\right)^2}{x^{(3)} - 2x^{(4)} + x^{(5)}};\ldots$$

(2) For turbulent flow in a smooth pipe, the friction factor f is given by the solution of von Karman's algebraic equation (Von Karman, 1934):

$$\sqrt{\frac{1}{f}} + 0.8 - 2 * \log_{10}\left(N_{Re} \cdot \sqrt{f}\right) = 0$$

where N_{Re} is the Reynolds number. Use a fixed-point method to find the value of f for the following values of the Reynolds number: $N_{\text{Re}} = 10^4, 10^5, 10^6$. A starting point for the coefficient of friction can be the Blasius formula: $f = 0.316(N_{\text{Re}})^{-0.25}$.

(3) We want to calculate the density of supercritical CO_2 at a pressure of 10^4 kPa and a temperature of 340 K. Under these conditions, an appropriate equation of state to characterize the p–v–T properties of the fluid is the Peng-Robinson equation of state (Lopez-Echeverry et al., 2017; Peng & Robinson, 1976) given by:

$$P = \frac{RT}{v - b} - \frac{a}{v(v + b) + b(v - b)}$$

where $a = 350$ m^{6}kPa/kmol^2 and $b = 0.07$ m^3/kmol.

Propose a fixed-point iteration formula and solve for the molar volume of the system.

(4) Carnaham and Starling (1969) proposed the following equation for the compressibility factor Z:

$$Z = \frac{1 + y + y^2 - y^3}{(1 - y)^3}; \; y = \frac{b}{4v}$$

where b is van der Waals' constant and v is molar volume. If $b = 0.08$ L/mol and $Z = 0.8$, propose a fixed-point iteration and find the molar volume of the system.

(5) The nonlinear equation $f(x) = e^x - 3x^2 = 0$ has three different solutions. The following fixed-point iteration function has been proposed:

$$\phi_1(x) = \pm\frac{e^{x/2}}{\sqrt{3}}$$

(a) The $\phi_1(x)$ function will converge to a root close to -0.5 if the negative sign is taken in the equation and will converge to a root close to 0.9 if the positive sign is taken. Why does this happen? Also, run three iterations to numerically check this in each case.

(b) The above equation is not useful to find the root close to the value 4. Why is this so? Propose a fixed-point iteration that converges to the root near the value 4 and perform three iterations to test the validity of your proposed method.

(c) Write a Python macro that executes a given number of fixed-point iterations (10 for example) for case (a), choosing one of two possible alternatives (positive root or negative root).

(6) Lee and Duffy (1976) proposed the following equation to calculate the friction factor f in the flow of a suspension of fibrous particles:

$$\frac{1}{\sqrt{f}} = \frac{1}{k}\ln(N_{\text{Re}}\sqrt{f}) + \left(14 - \frac{5.6}{k}\right)$$

where N_{Re} is the Reynolds number and k is a parameter that depends on the concentration of the suspension. If $N_{\text{Re}} = 4000$ and $k = 2.8$, write a script in Python that solves the above equation for f using the fixed-point method, from the initial estimate $f = 0.001$. The macro must decide at what point to stop the fixed-point iteration, using an appropriate error criterion for the variable f, defined by you.

(7) The Ergun equation (Ergun, 1952; Quinn, 2014) allows you to calculate the pressure drop of a fluid circulating through a packed bed:

$$\frac{\Delta P}{L} = 150\left(\frac{\eta v_0}{D_{\text{P}}^2}\right)\frac{(1-\varepsilon)^2}{\varepsilon^3} + 1.75\left(\frac{\rho v_0^2}{D_{\text{P}}}\right)\left(\frac{1-\varepsilon}{\varepsilon^3}\right)$$

where ΔP is the pressure drop, L is the height of the solids bed, η is the viscosity and v_0 the surface velocity of the fluid, D_{P} is the particle size, ρ the density of the fluid, and ε is the porosity of the bed ($0 < \varepsilon < 1$). Program a Python script that calculates porosity ε when all other variables are known and uses the fixed-point method with tol $= 10^{-6}$, maxit $= 100$. Test your script with the following data (all SI units consistent with each other):

$$\eta = 0.01 \text{ Poise},\ v_0 = 0.01 \text{ m/s},\ D_{\text{P}} = 0.049 \text{ m},$$
$$\rho = 2000 \text{ kg/m}^3,\ L = 1.6 \text{ m and } \Delta P = 416 \text{ Pa}.$$

(8) A free-falling particle reaches a terminal velocity v in m/s which is given by the following algebraic equation:

$$1.15v^2 + 1.4v^{1.5} = 1500$$

(a) Solve this equation using Newton's method.
(b) Set a convergence criterion to stop iteration and incorporate it into a Python script.
(c) Check whether you got convergence, whether the method works or not, etc.

(9) The nonlinear equation

$$f(x) = e^x - 3x^2 = 0$$

has three different solutions in the interval $[-1, 4]$. Program a script in Python that solves the above equation using Newton's method for different initial values $x^{(0)}$ then apply it to find the three roots of $f(x)$.

(10) Solve the above problem by using the quadratic interpolation method to find the three roots of the algebraic equation.

(11) For a chemical equilibrium problem, the following nonlinear equation must be solved:

$$f(x) = 3.1x\sqrt{1+x} - (1-x)\sqrt{10.5+x} = 0$$

where $0 < x < 1$. Write a Python script that uses the quadratic interpolation method to solve this equation. How might you decide whether your iteration is converging or not? Explain.

(12) For turbulent flow in a pipe with diameter D and roughness e, the friction factor f is given by the solution of the equation of Colebrook and White (1937):

$$\frac{1}{\sqrt{f}} = 1.14 - 0.85 \cdot \ln\left(\frac{e}{D} + \frac{9.35}{N_{Re}\sqrt{f}}\right)$$

where N_{Re} is Reynolds number.

(a) Program a script in Python to find f that takes the values of N_{Re} and e/D as arguments and finds f using Newton's method.
(b) Now make a script in Python that calculates the value of f for different combinations $N_{\mathrm{Re}} = 10^4$ to 10^8, $e/D = 0.0001$ to 0.01, and then makes a plot that represents the function $f(N_{\mathrm{Re}}, e/D)$ in the domain where the calculations were made. What is the name of this plot?

(13) Consider the sedimentation of a solid spherical particle in a liquid medium at rest. The particle is subjected to the force of gravity (F_g) and the drag force with the liquid (F_d). The equation for the terminal velocity V_t of a particle of diameter D_P is given by the expression:

$$f(V_t) = 3 \cdot \rho \cdot C_D \cdot (V_t)^2 - 4 \cdot g \cdot (\rho_P - \rho) \cdot D_P = 0$$

where ρ and ρ_P are the densities of the liquid and the particle, respectively; g is the acceleration of gravity (9.8 m/s^2); and the friction factor C_D is determined by the Reynolds number of the particle: $\mathrm{Re} = V_t \cdot \rho \cdot D_P/\mu$, ($\mu$ is the viscosity of the liquid) by the following expression (Morrison, 2013):

$$C_D = \frac{24}{\mathrm{Re}} + \frac{2.6\left(\frac{\mathrm{Re}}{5}\right)}{1+\left(\frac{\mathrm{Re}}{5}\right)^{1.52}} + \frac{0.411\left(\frac{\mathrm{Re}}{2.63\times10^5}\right)^{-7.94}}{1+\left(\frac{\mathrm{Re}}{2.63\times10^5}\right)^{-8}} + \frac{0.25\left(\frac{\mathrm{Re}}{10^6}\right)}{1+\left(\frac{\mathrm{Re}}{10^6}\right)}.$$

Program a Python script that receives input arguments ρ, μ, ρ_P, D_P, and deliver as exit arguments: (a) Reynolds number Re, (b) coefficient of friction C_D, and (c) the sedimentation rate V_t (in m/s). Test your function with the following input data:

(i) $\rho = 1000$ kg/m^3, $\mu = 9\times10^{-4}$ kg/[m s], $\rho_P = 2600$ kg/m^3, $D_P = 0.02$ m;
(ii) $\rho = 1000$ kg/m^3, $\mu = 9\times10^{-4}$ kg/[m s], $\rho_P = 2000$ kg/m^3, $D_P = 2\times 10^{-4}$ m;
(iii) $\rho = 1000$ kg/m^3, $\mu = 9\times10^{-4}$ kg/[m s], $\rho_P = 1500$ kg/m^3, $D_P = 2.5\times 10^{-6}$ m;

(14) It is necessary to calculate the specific volume (x) of a compressed gas, which is obtained by solving the following cubic equation in the interval [0 1]:

$$f(x) = x^3 - 0.068x^2 + 0.0005x - 0.24 = 0.$$

(a) Explain why Newton's method is not advisable to be applied in this type of equation.
(b) Check if it is feasible to use the secant method to solve this problem.

(15) Consider a liquid mixture of benzene (B) and toluene (T) at a total pressure of 760 mmHg. When the temperature of the mixture increases at constant pressure, the first vapor bubble that is produced corresponds to the bubble point of the mixture and the respective temperature at which it occurs corresponds to the bubble temperature of the binary mixture.

The equation that holds at the bubble point is that the vapor pressure of the binary mixture is equal to the total pressure. Assuming ideal liquid solution, then it is true that:

$$P_{\text{Total}} = x_B P_B(T) + (1 - x_B) P_T(T)$$

where x_B is the molar fraction of benzene; P_B and P_T are the saturation pressures of both compounds, which are given by the following equations:

$$P_B = \exp\{17.5318 - 3848.1/(273.15 + T(°\text{C}))\}$$
$$P_T = \exp\{17.913 - 4328.1/(273.15 + T(°\text{C}))\}.$$

Construct a Python script that evaluates the bubble temperature (°C) for binary mixtures with $x_B = 0.1, 0.2, \ldots, 0.9$ and plots that temperature as a function of x_B, including axis titles, etc.

(16) In Shacham (1988) there are numerous examples of nonlinear equations that arise in different areas of Chemical Process Engineering. Using an absolute tolerance of 10^{-8}, solve the problems listed in Tables 2 and 3 of that paper and compare your results with those of the paper for the cases of the secant, Newton, fixed-point, and Wegstein methods.

(17) Chemical equilibrium problems in mixtures of compounds produce systems of nonlinear equations with multiple solutions. Consider, for example, the following system of equilibrium reactions

$$3H_2 + CO \leftrightarrow CH_4 + H_2O;$$
$$CO + H_2O \leftrightarrow CO_2 + H_2$$
$$CO_2 + C \leftrightarrow 2CO$$
$$5H_2 + 2CO \leftrightarrow C_2H_6 + 2H_2O$$

If 3 moles of H_2 are initially combined with 1 mole of CO until equilibrium is reached, equilibrium concentrations are obtained by solving the following system of algebraic equations:

$$\begin{array}{ll} K_1(A)^3B - x_1C(N)^2 = 0 & A = 3 - 3x_1 + x_2 - 5x_4 \\ K_2BC - (x_2 - x_3)A = 0 & B = 1 - x_1 - x_2 + 2x_3 - 2x_4 \\ K_3(x_2 - x_3)N - (B)^2 = 0 & C = x_1 - x_2 + 2x_4 \\ K_4(A)^5(B)^2 - x_4(CN)^2 = 0 & N = 4 - 2x_1 + x_3 - 4x_4 \end{array}$$

Find a solution for chemical equilibrium in this example (molar percentages of each gas), if the equilibrium constants are given by: $K_1 = 70, K_2 = 5, K_3 = 0.01, K_4 = 0.1$.

The moles of gases in equilibrium are given by the expressions:

$$\begin{array}{ll} H_2 : 3 - 3x_1 + x_2 - 5x_4 & CH_4 : x_1 \\ CO : 1 - x_1 - x_2 + 2x_3 - 2x_4 & CO_2 : x_2 - x_3 . \\ H_2O : x_1 - x_2 + 2x_4 & C_2H_6 : x_4 \end{array}$$

Here x_1, ..., x_4 represent the degrees of reaction advance in each of the equilibrium equations.

Construct a script that solves this problem by using the *Python fsolve* function. Your script should also verify if the solution found satisfies the criteria of delivering positive mole numbers for each compound. Compare the efficiency of convergence of *fsolve* by solving this problem with a discrete Newton method for this system of equations.

(18) Consider a coal gasification process that aims to produce a fuel gas, which is a mixture of CO, CO_2, and hydrogen (H_2). To do this, water vapor is added to a bed of biomass at a high temperature (about 500 °C) and the following system of chemical reactions in equilibrium is obtained.

$$\begin{array}{l} C + 2H_2O \leftrightarrow CO_2 + 2H_2 \\ C + H_2O \leftrightarrow CO + H_2 \\ C + 2H_2 \leftrightarrow CH_4 \end{array} .$$

If 1 mole of H_2O is initially combined with biomass until equilibrium is reached, equilibrium concentrations are obtained by solving the following system of algebraic equations:

$$
\begin{aligned}
&K_1(B)^2C - x_1(A)^2 = 0 \quad && A = 2x_1 + x_2 - 2x_3 \\
&K_2BC - x_2A = 0 && B = 1 - 2x_1 - x_2 \\
&K_3(A)^2 - x_3C = 0 && C = 1 + x_1 + x_2 - x_3.
\end{aligned}
$$

Find a solution for chemical equilibrium in this example (molar percentages of each gas), if the equilibrium constants are given by: $K_1 = 0.1,\ K_2 = 0.05,\ K_3 = 2$.

The moles of gases in equilibrium are given by the expressions:

$$
\begin{aligned}
&H_2 : 2x_1 + x_2 - 2x_3 \quad && CH_4 : x_3 \\
&CO : x_2 && CO_2 : x_1 \\
&H_2O : 1 - 2x_1 - x_2 &&
\end{aligned}
$$

Here x_1, ..., x_3 represent the degrees of reaction advance in each of the equilibrium equations.

Solve this problem by using Newton's discrete method. To verify that the numerical solution obtained is correct, use the expressions for the numbers of moles to decide which is the physical solution of the problem (it is the one that should produce only positive numbers of moles).

(19) This problem arises from the optimization of two variable functions:

$$
f_1(x_1, x_2) = -\frac{1}{(x_1)^2} + \left(\frac{1-x_2}{x_2} + \frac{1}{1-x_2}\right)\frac{1}{(1-x_1)^2} = 0
$$

$$
f_2(x_1, x_2) = -1 + \frac{1}{(1-x_1)}\left(\frac{2x_2 - 1}{(1-x_2)^2}\right) = 0.
$$

(a) Solve this system by using Newton's method with $x^{(0)} = [0.5\ 0.5]'$. Approximate the Jacobian by finite differences.

(b) Now solve the problem by iterating the fixed point so that now the equations are written as follows:

$$
x_1 = (1 - x_1)\left(\frac{1-x_2}{x_2} + \frac{1}{1-x_2}\right)^{-0.5} = g_1(x_1, x_2)
$$

$$
x_2 = \frac{(1-x_1)(1-x_2)^2 + 1}{2} = g_2(x_1, x_2).
$$

(20) We want to solve the system of equations:

$$-13.6x_1 + 20.4x_2 - 6.8x_3 = \Phi \cdot \left(\frac{x_1}{1 + K \cdot x_1}\right)$$

$$14.6x_1 - 91.4x_2 + 76.8x_3 = \Phi \cdot \left(\frac{x_2}{1 + K \cdot x_2}\right)$$

$$0.95x_1 - 14.95x_2 + 14x_3 = Bi(1 - x_3).$$

That represents the concentration inside a porous solid in which the degradation of a chemical compound is occurring. The unknowns (x_1, x_2, x_3) are the values of the concentrations of the compound in different spatial locations within the solid.

Construct a Python script that receives the values of the K and Bi parameters as an input argument and compute the solution of the system of equations above for the following parameter values Φ: 0.01, 0.1, 1, 10, 100, and 1000, using Newton's method.

It should be checked that the number of iterations (for each value of Φ) does not exceed a maximum value of 50 iterations, and as an error criterion use the error: err $= ||f(x^{(k)})||$, which must be less than 10^{-6} to consider the iteration successful.

At the end of the function, a graph must be delivered, indicating the values of (x_1, x_2, x_3) depending on the parameter Φ. Test your script with $K = 0.1$, 1, 10 and Bi $= 1$.

(21) Consider an instantaneous evaporator (or flash) into which a feed stream (F moles/h) consisting of three components is entered, which is instantly separated into a vapor stream (V moles/h) and a liquid stream (L moles/h). The balance of moles for each component (steady-state) can be set as

$$z_1 = x_1[1 + \alpha(k_1 - 1)]; \quad z_2 = x_2[1 + \alpha(k_2 - 1)]; \quad z_3 = x_3[1 + \alpha(k_3 - 1)]$$

where z_i is the molar fraction of the component i in the feed, $\alpha = V/F$ is the feed fraction that evaporates (on a molar basis) and k_i is the liquid–vapor equilibrium constant for the component i:

$$y_1 = k_1x_1; \quad y_2 = k_2x_2; \quad y_3 = k_3x_3$$

where the coefficients k_i are given by the expression: $k_i - P_i^S/P$, where P_i^S is the vapor pressure of the pure compound, and P is the total pressure of the system (ideal behavior in the liquid phase is assumed). To find the vapor pressures of pure components, Antoine's equation is used:

$$\log_{10}\left(P_i^S\right) = A_i - \frac{B_i}{(C_i + t)} \quad i = 1, 2, 3$$

where A_i, B_i, and C_i are constant for each compound and t is the temperature of the system in °C, and $P_i{}^S$ measured in mmHg. For this system, the constants are given by the values: $A_1 = 8.1, B_1 = 1400, C_1 = 200; A_2 = 7.7, B_2 = 1720, C_2 = 215, A_3 = 8.3, B_1 = 1250, C_1 = 195$.

It must be remembered that the restrictions on compositions in the vapor and liquid must be complied with:

$$\sum x_i = 1; \quad \sum y_i = 1.$$

Propose a problem-solving scheme, and calculate the composition of the liquid and vapor, the pressure and temperature of the system, and the flows V and L if it is known that $F = 100$ mol/h, and that $Z = [0.1\ 0.5\ 0.4]$. Your script should also explicitly verify that the solution found satisfies the design equations of the flash evaporator.

(22) The following first-order elemental reactions occur in the liquid phase in a chemical reactor:

$$2A \overset{k_1}{\rightarrow} B \overset{k_2}{\rightarrow} C$$

The kinetics constants are: $k_1 = 1.0$ [l/mol min] and $k_2 = 0.1$ [1/min]. Consider that the concentration of input for compound A is $C_{A0} = 1$ [mol/L].

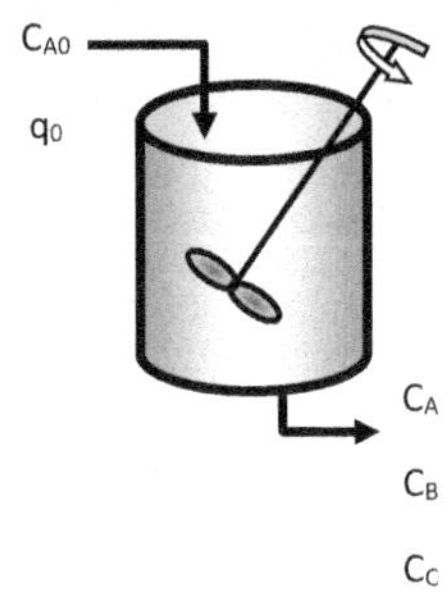

(a) Using Newton's method, calculate the concentrations of compounds A, B, and C at steady state for a residence time (τ) equal to 0.5 min.

(b) Graph the concentration of compound B as a function of residence time τ for a range of 0.01–30 min. Determine the maximum value that compound B reaches. For each τ compare your solution with the one that delivers the *fsolve* function.

The equations of the physical model to be solved are the following:

$$\frac{C_{A0} - C_A}{\tau} - 2 \cdot k_1 \cdot C_A^2 = 0$$

$$\frac{-C_B}{\tau} + k_1 \cdot C_A^2 - k_2 \cdot C_B = 0$$

$$\frac{-C_C}{\tau} + k_2 \cdot C_B = 0$$

(23) Consider the following system of nonlinear equations:

$$\begin{aligned} f_1(x_1, x_2) &= x_1 - 4 \cdot (x_1)^2 - x_1 \cdot x_2 = 0 \\ f_2(x_1, x_2) &= 2 \cdot x_2 - (x_2)^2 + 3 \cdot x_1 \cdot x_2 = 0 \end{aligned}$$

Find all the solutions to this system of equations using the Newton method. If only solutions with values greater than and equal to zero are accepted, how many valid solutions are there? If only solutions with values between 0 and 1 are accepted, how many valid solutions are there?

(24) Consider the following fluid mechanics problem, where a main pipe (1–2) forks into two runs: (2–3) and (2–4). The equations for the pressure loss in each section plus the conservation of mass produce the following system of equations:

$$\begin{aligned} p_1 - p_2 &= K_1 \cdot (Q_1)^{1.75} \\ p_2 - p_3 &= K_2 \cdot (Q_2)^{1.75} \\ p_2 - p_4 &= K_3 \cdot (Q_3)^{1.75} \\ Q_1 &= Q_2 + Q_3 \end{aligned}$$

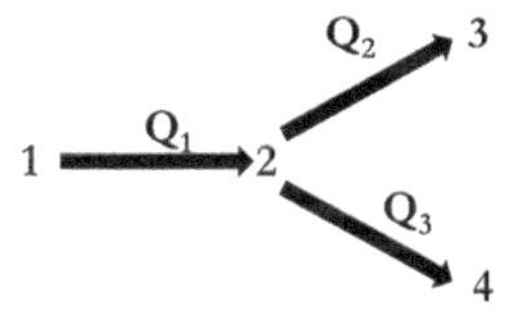

where pressures are in psi and flow rates in gpm. The following data are known:

$$\begin{aligned} &p_1 = 75 \text{ psi}, p_3 = 20 \text{ psi}, \ p_4 = 15 \text{ psi}, \\ &K_1 = 3.25\text{e} - 3, K_2 = 4.67\text{e} - 3, \ K_3 = 3.72\text{e} - 2. \end{aligned}$$

Compute the values of Q_1, Q_2, Q_3, and p_2 using the discrete Newton method.

(25) In the analysis of the stresses to which the turbine of an aircraft is subjected, it is necessary to solve the following set of nonlinear equations:

$$\begin{aligned} &3 + \frac{2}{r^2} - \frac{1}{8}\frac{(3-2v)}{(1-v)}\omega^2 r^2 = 4.5 \\ &6v - \frac{1}{2}\frac{v}{(1-v)}\omega^2 r^2 = 2.5 \\ &3 - \frac{2}{r^2} - \frac{1}{8}\frac{(1+2v)}{(1-v)}\omega^2 r^2 = 0.5 \end{aligned}$$

Solve the equations to find r, v, and ω, using the *fsolve* function. Also explicitly display the iterations and intermediate results.

References

Aitken, A. (1926). On Bernoulli's numerical solution of algebraic equations. *Proceedings of the Royal Society of Edinburgh, 46*, 289–305.

Carnahan, N. F., & Starling, K. E. (1969). Equation of state for nonattracting rigid spheres. *The Journal of Chemical Physics, 51*, 635–636.

Colebrook, C. F., & White, C. M. (1937). Experiments with fluid friction in roughened pipes. *Proceedings of the Royal Society of London Series A Mathematical and Physical Sciences, 161*(906), 367–381.

Dennis, J. E., & Mori, J. J. (1977). Quasi-Newton methods, motivation and theory. *SIAM Review, 19*, 46–89.

Ergun, S. (1952). Fluid flow through packed column. *Chemical Engineering Progress, 49*, 89–94.

Lee, P. F. W., & Duffy, G. G. (1976). Relationship between velocity profiles and drag reduction in turbulent fiber suspension flow. *AIChE Journal, 22*, 750–753.

Lopez-Echeverry, J. S., Reif-Acherman, S., & Araujo-Lopez, E. (2017). Peng-Robinson equation of state: 40 years through cubics. *Fluid Phase Equilibria, 447*, 39–71.

Morrison, F. A. (2013). *An introduction to fluid mechanics*. Cambridge University Press.

Peng, D. Y., & Robinson, D. B. (1976). A new two-constant equation of state. *Industrial and Engineering Chemistry Fundamentals, 15*, 59–64.

Quinn, H. M. (2014). A reconciliation of packed column permeability data: Deconvoluting the Ergun papers. *Journal of Materials, 2014*, 548482. https://doi.org/10.1155/2014/548482

Sargent, R. W. H. (1981) A review of methods for solving non-linear algebraic equations. In *Foundations of computer-aided chemical process design*. Engineering Foundation.

Shacham, M. (1988). An improved memory method for the solution of a nonlinear equation. *Chemical Engineering Science, 44*(7), 1495–1501.

von Karman, T. (1934). Turbulence and skin friction. *Journal of the Aeronautical Sciences, 1*(1), 1–20.

Chapter 3
Ordinary Differential Equations

Many models encountered in practice are dynamic, in that the system evolves over time from a given initial condition. Indeed, a process almost never operates under steady-state conditions, making it relevant to understand the dynamic behavior of the system.

Incorporating more realistic assumptions into the models involves considering nonlinear terms, non-conventional forcing functions, etc., which prevents analytical solutions to the model equations. Therefore, we must seek efficient numerical methods to solve these cases with the aid of a computer.

Most dynamic models can be formulated as ordinary differential equations (ODE) of the form:

$$F(x, y, y', y'', \ldots, y^n) = 0. \tag{3.1}$$

Typically, boundary or initial conditions specify the values of y or its higher-order derivatives at specified values of x. If all these conditions are specified at a common point x_0, then Eq. (3.1) is called an initial value problem (IVP); otherwise, it is a boundary value problem (BVP). In this chapter, we devote our attention to the IVP case.

An nth-order equation such as Eq. (3.1) can always be written as a system of n first-order ODE (by defining $n - 1$ new dependent variables). Consequently, we only need to consider the solution of systems of first-order ODE. Today there exists a substantial amount of general-purpose software for systems of first-order ODE of the form:

$$y_i'(x) = f_i(x, y_1, y_2, \ldots, y_n); \quad y_i(x_0) = y_{i,0}; \quad i = 1, 2, \ldots, n \tag{3.2}$$

And these are the ones typically considered in this chapter.

H. Jorquera and C. A. Gelmi, *Numerical Methods in Chemical Process Engineering Using Python*, https://doi.org/10.1007/978-3-032-22958-8_3

3.1 How Do Numerical Methods Operate?

Since the solution of an IVP such as Eq. (3.2) is (in principle) no more difficult than solving a single first-order ODE, it is sufficient to consider algorithms for solving the scalar equation:

$$y'(x) = f(x, y); \quad y(x_0) = y_0. \tag{3.3}$$

All algorithms that solve Eq. (3.3) are discrete methods that start from the initial condition x_0 and generate approximations y_i to the exact solution $y(x_i)$ at discrete values of the independent variable, x_i, with

$$x_{i+1} = x_i + h_i \tag{3.4}$$

where h_i is the step size in the i-th increment. It is important to emphasize that the numerical methods we present do not deliver the exact solution $y(x)$, which is a function defined for any value of x in the domain of the function y, but rather a discrete set of approximations $\{y_i\}$. This is because a computer cannot work with an infinite number of x-values; it must select a finite set of points $\{x_i\}$ at which the numerical solution is estimated, and this selection is summarized in Eq. 3.4. Sometimes the steps h_i are identical, and then the coordinates of the independent variable are:

$$x_i = x_0 + i \cdot h_i; \quad i = 1, 2, \ldots, n \tag{3.5}$$

where the global integration interval is $[x_0, x_n]$, as shown in Fig. 3.1.

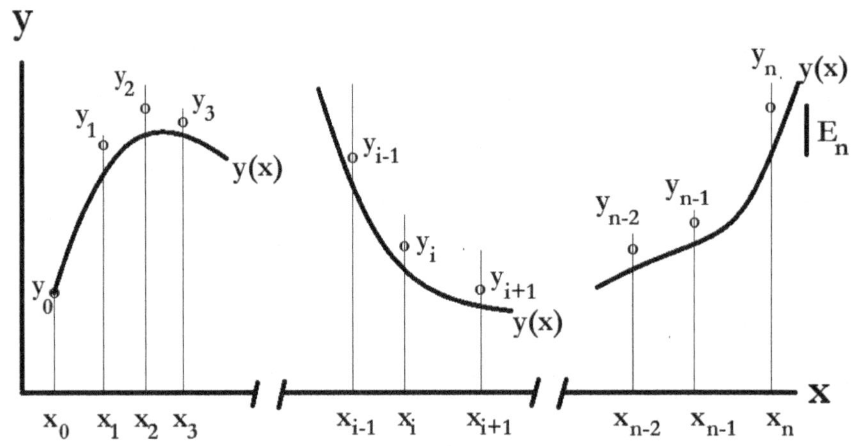

Fig. 3.1 Numerical and exact solution on the interval $[x_0, x_n]$. The solid curve is the exact solution $y(x)$, and the circles represent the approximations produced by a numerical integration method

In the absence of rounding error, the difference between y_i and $y(x_i)$, that is, between the computed approximation and the exact solution, is called the discretization error:

$$E_i = y_i - y(x_i) \tag{3.6}$$

E_i is determined solely by the nature of the approximations in the method. The change in this error when moving from x_i to x_{i+1} is composed of a propagated error and a local truncation error. The local truncation error arises because the method cannot solve the differential equation exactly and must make some assumption to estimate it (for example, a truncated Taylor series expansion). Moreover, as the solution is advanced across the points x_i, the method works not only with an approximate equation in the algorithm but also inputs the estimated values y_i rather than the exact $y(x_i)$, which gives rise to error propagation.

If the integration step h is small, the change in E_i at each step is dominated by the local truncation error, which has the form $K \cdot h^{p+1}$ (K depends on certain derivatives of f). The local truncation error is said to be of order $(p+1)$ [$O(h^{p+1})$], and the algorithm is called a pth-order method; this implies that for a solution of the form $y(x) = x^p$, the local truncation error is zero, or in other words, the global error accumulated at the end of the integration is $O(h^p)$.

3.2 One-Step Methods

In all these methods, the solution at x_{i+1} is obtained exclusively from the information available at x_i. The simplest (and oldest) algorithms for the solution of IVPs are attributed to Euler. His explicit algorithm is given by:

$$y_{i+1} = y_i + h \cdot f(x_i, y_i). \tag{3.7}$$

With a local truncation error of $f'(\xi, y(\xi)) \cdot h^2/2$, where $x_i < \xi < x_{i+1}$. This follows directly from the Taylor expansion of $y(x)$ about x_i. In the absence of rounding, the global truncation error at x_{i+1} is:

$$E_{i+1} = E_i \cdot \left\{1 + h \cdot \left(\frac{\partial f}{\partial y}\right)(x, \alpha)\right\} - \frac{h^2}{2} f'(\xi, y(\xi)) \tag{3.8}$$

where α is in $[y_i, y(x_i)]$ and ξ is in $[x_i, x_{i+1}]$. The first term on the right-hand side of Eq. (3.8) corresponds to the propagation of errors in y_i, and the second term is the local truncation error. The global truncation error is $O(h)$, that is, Euler is a first-order method. The Euler method has geometric interpretation shown in Fig. 3.2.

In the explicit Euler method, the integral is approximated by

Fig. 3.2 Euler's method

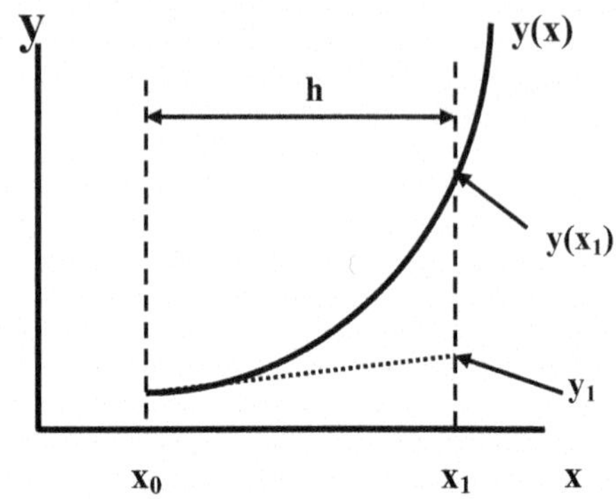

$$y_{i+1} = y(x_i) + h \int_{x_i}^{x_i+1} f(s, y(s))\mathrm{d}s \tag{3.9}$$

setting that $f(x, y(x))$ is constant over the interval $[x_i, x_{i+1}]$ and has the value $f(x_i, y_i)$. For many ODE, the Euler method is not sufficiently accurate, i.e., it requires many steps (evaluations of $f(x, y(x))$) to solve an ODE over a given interval with acceptable accuracy.

3.2.1 *Explicit Runge–Kutta Methods*

An alternative approach to improve the accuracy of the Euler method would be to include more terms in the Taylor series expansion:

$$y(x_{i+1}) = y(x_i) + h \cdot f(x_i, y(x_i)) + \frac{h^2}{2} \cdot f'(x_i, y(x_i)) + \cdots \tag{3.10}$$

Unfortunately, this becomes impractical for most ODE, since the derivatives of f must be determined using the chain rule, which requires symbolic manipulation. Moreover, evaluating f typically involves expensive expressions; this motivates usage only evaluations of f, and as few as possible, during the numerical integration.

To overcome the limitation posed by the low accuracy of the Euler method, Runge (1895), Heun (1900), and Kutta (1901) proposed methods that require only the computation of y' (that is, $f(x, y)$) but that achieve accuracy comparable to what would be obtained from Eq. (3.10) if more than two terms were retained in the series. All these methods are explicit in y_{i+1} and have the following generic form:

$$y_{i+1} = y_i + h \cdot \phi(x_i, y_i, h) \tag{3.11}$$

where ϕ is the increment function, which corresponds to a weighted average of the values of f computed at r points in the interval $[x_i, x_{i+1}]$. For Runge–Kutta methods of order less than or equal to five, a method of order p requires exactly p evaluations of f over the interval. For example, the explicit Runge–Kutta methods of second order have the form:

$$\begin{aligned} y_{i+1} &= y_i + a \cdot k_1 + b \cdot k_2; \\ k_1 &= h \cdot f(x_i, y_i); \\ k_2 &= h \cdot f(x_i + p \cdot h, y_i + q \cdot k_1); \end{aligned} \tag{3.12}$$

The parameters a, b, p, and q are determined by expanding k_2 in a Taylor series, substituting k_1 and k_2, and enforcing consistency with the Taylor expansion of $y(x)$ (Eq. 3.10) for the terms h^0, h^1, and h^2. This yields three equations for four parameters, which means that one of them can be fixed arbitrarily. In this way, a family of second-order Runge–Kutta methods is obtained.

Analogously, higher-order Runge–Kutta methods can be generated. The most widely used are the explicit fourth-order Runge–Kutta methods attributed to Runge (1895) and Gill (1951). There are also higher-order methods that are explicit. In most cases, methods of order greater than four require more than p evaluations of f on $[x_i, x_{i+1}]$. Among these methods are the fifth- and sixth-order methods attributed to Butcher (1963), Luther and Konen (1955), Luther (1966, 1968), and Felhberg (1968).

3.2.2 *Local Truncation Error and Its Control During Numerical Integration*

The local truncation error for an order-p Runge–Kutta method arises from the truncated Taylor series that produced it, and has the form

$$E_t = K \cdot h^{p+1} + O\left(h^{p+2}\right) \tag{3.13}$$

where K is a complicated function of f and its derivatives, i.e., of the exact solution to the initial value problem, and is therefore incomputable.

Thus, we face two challenges: (i) estimating the value of K and (ii) ensuring that the error E_t in Eq. (3.13) does not exceed a prescribed tolerance.

The solution to these two challenges is obtained by allowing the integration step h to vary, using estimations of the local truncation error E_t produced at each integration step (Butcher, 2007). To accomplish this, one works with pairs of Runge–Kutta methods that are integrated in parallel, one of order p and another of a higher order q, with the same step size at each stage; typically $q = p + 1$ is chosen to optimize computation time. Then, the difference between the results of each method separately (Δy) provides an asymptotically correct estimator of the error E_t. For small values of h, the difference behaves asymptotically as $h^{(q+1)}$, and h can be controlled so that that difference remains bounded. For example, if ε is the acceptable tolerance for the local truncation error, the new integration step is given by:

$$h_{\text{new}} \leq \left(\frac{\varepsilon}{\| \Delta y_{\text{current}} \|} \right)^{1/(q+1)} \cdot h_{\text{current}}. \tag{3.14}$$

In Python, $2^{\text{nd}}/3^{\text{rd}}$-order and $4^{\text{th}}/5^{\text{th}}$-order, explicit methods are implemented in the *solve_ivp* routine, from package SciPy, by setting the argument "method" equal to "RK23" and "RK45", respectively. The former implements a (2, 3) pair of explicit Runge–Kutta methods developed by Bogacki and Shampine (1989), while the latter uses a (4, 5) pair proposed by Dormand and Prince (1980).

It is recommended to first test the numerical resolution of an IVP using either of these two Python-implemented methods; if the numerical solution is slow, this indicates stability-related issues that require the use of specialized techniques, which will be presented in Sects. 3.4 and 3.5.

3.2.3 Implicit Runge–Kutta Methods

When the update function ϕ in Eq. (3.11) includes y_{i+1} in the evaluation, the method is implicit, and to obtain y_{i+1} one must solve a nonlinear equation, as discussed in Chap. 2. In that case, we are dealing with implicit Runge–Kutta methods, and the weighting factors in Eq. (3.11) correspond to the quadrature coefficients of Gauss–Legendre, Radau, or Lobatto methods, among the most common (Butcher, 2007). Butcher (1964), Caillaud and Padmanabhan (1971), Rosenbrok (1963) and Michelsen (1976) provide greater detail on how these implicit methods are constructed. In general, these methods are not widely used because they do not typically compete in precision and efficiency with the linear multistep methods discussed in Sect. 3.3.

In Python, the implicit method 'Radau' is available within *solve_ivp*, corresponding to an implicit Runge–Kutta method developed by Shampine and Hosea (1996). This function can be applied when the methods recommended by default, RK23 and RK45, take too long to deliver a solution.

3.2.4 Conclusions Regarding Runge–Kutta Methods

Explicit Runge–Kutta methods have proven efficient for most first-order ODE systems, and nearly every software library includes a general-purpose RK routine. RK methods offer the following advantages:

1. They require only the differential equation and the initial condition to start.

2. Since each new step can be regarded as the beginning of a new IVP, the integration step can be readily adjusted for subsequent applications of the method.
3. No iterative solution is required at each step of the integration (except for implicit methods, which are used rarely).
4. The methods are easy to implement and require little memory to store past information (previous y_i and f_i values).
5. They produce highly accurate results, often better than those obtained with other methods at comparable step sizes.

The main disadvantages of RK methods are:

1. It is difficult to estimate the local truncation error and, therefore, to establish automatic step-size control criteria. The standard remedy is to apply methods in pairs (RK23, RK45 in Python).
2. Some other methods require fewer evaluations of f per step. For example, a fourth-order RK method typically requires about twice as many evaluations of f as a linear multistep method of the same order or accuracy.

3.3 Linear Multistep Methods (LMM)

The second class of methods aimed at solving IVPs for ODE are linear multistep methods (LMMs, or multistep methods), which possess the following generic form:

$$y_{i+1} = \sum_{k=0}^{p} \alpha_k y_{i-k} + h \cdot \sum_{k=-1}^{p} \beta_k f_{i-k} \tag{3.15}$$

where $f_{i-j} = f(x_{i-j}, y_{i-j})$. These methods typically do not have an initiation, and it is necessary to use a one-step method (e.g., Runge–Kutta) to generate the required values of f and y before the Eq. (3.15) can be applied for the first time. When $\beta_{-1} = 0$, Eq. (3.15) is explicit in y_{i+1}, and the algorithm is called explicit or predictor method. When $\beta_{-1} \neq 0$, the method is implicit in y_{i+1} and the algorithm is called implicit or corrector method; in such a case, Eq. (3.15) must be solved by some iterative method, as discussed in Chap. 2.

Some of the most common closed- and open-type algorithms can be obtained by fitting the points f_{i-j} ($j = 1, 2, \ldots, k$) with an interpolating polynomial and then evaluating y_{i+1} with the expression

$$y_{i+1} = y_{i-q} + \int_{x(i-q)}^{x(i+1)} \theta_r(x)\mathrm{d}x \tag{3.16}$$

where $\theta_r(x)$ is an interpolating polynomial of degree r (Nakamura, Chapter 2). Given the values of q and r, this yields algorithms that are essentially Newton–Cotes quadrature formulas (Rice & Do, 2012).

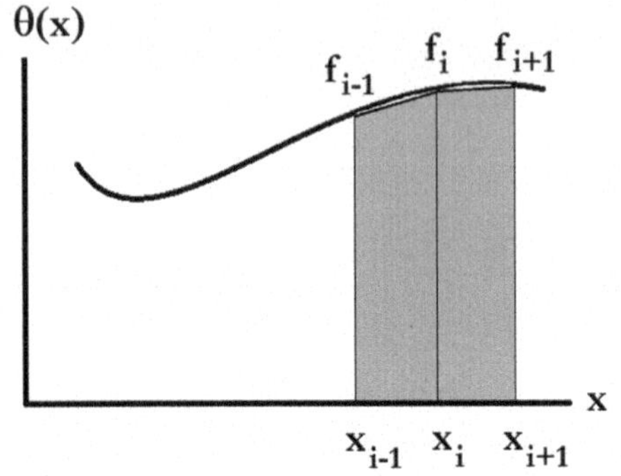

Fig. 3.3 Integral of Eq. (3.16) for Simpson's rule

Example 3.1 Milne's method

Milne's fourth-order predictor–corrector method is given by:

$$P : y_{i+1} = y_{i-3} + \frac{4}{3} \cdot h \cdot (2f_i - f_{i-1} + 2f_{i-2});\ E_t = \frac{14}{45} \cdot h^5 \cdot f^{(5)}(x*)$$

$$C : y_{i+1} = y_{i-1} + \frac{1}{3} \cdot h \cdot (f_{i+1} + 4f_i + f_{i-1});\ E_t = -\frac{1}{90} \cdot h^5 \cdot f^{(5)}(x*)$$

And the expressions for the local truncation error come from the respective Newton–Cotes formulas (Nakamura, 1991). The equation for the corrector is Simpson's quadrature rule. Figure 3.3 shows the shaded area, which is the integral in Eq. (3.16) evaluated using Simpson's rule.

There is no Python version of this method, because Milne's method has numerical stability issues that will be discussed below. The reader is encouraged to program this method to check its performance though.

3.3.1 Construction of LMM

The simplest and most general way to obtain LMM algorithms is to choose the order p of the method; then the coefficients α_j, $\beta_j \neq 0$ are determined by requiring that Eq. (3.15) be exact for polynomials of degree k or less, which yields the system of equations:

$$\sum_{k=0}^{p} \alpha_k = 1;\quad \sum_{k=0}^{p} (-k)^j \alpha_k + j \cdot \sum_{k=-1}^{p} (-k)^{j-1} \beta_k = 1;\quad j = 1, 2, \ldots, k. \tag{3.17}$$

Since there are more coefficients $(2p+3)$ than equations $(k+1)$, some coefficients must be chosen and fixed, which gives rise to the various MLMs proposed in the literature.

Example 3.2 Derivation of Milne's method

The Milne corrector method can be derived by requiring that the expression:

$$y_{i+1} = \alpha_1 y_{i-1} + h \cdot (\beta_{-1} f_{i+1} + \beta_0 f_i + \beta_1 f_{i-1} + \beta_2 f_{i-2})$$

Produce the correct solution for functions of the form $y(x) = 1, x, x^2, x^3$, and x^4; applying Eq. (3.17) yields a system of five linear equations for five unknowns:

$$\begin{aligned}
&\alpha_0 + \alpha_1 + \alpha_2 + \alpha_3 + \alpha_4 = 1 \rightarrow \alpha_1 = 1 \\
&-\alpha_1 + \beta_{-1} + \beta_0 + \beta_1 + \beta_2 = 1 \rightarrow \beta_{-1} + \beta_0 + \beta_1 + \beta_2 = 2 \\
&\alpha_1 + 2 \cdot \{\beta_{-1} - \beta_1 - 2\beta_2\} = 1 \rightarrow \beta_{-1} - \beta_1 - 2\beta_2 = 0 \\
&-\alpha_1 + 3 \cdot \{\beta_{-1} + \beta_1 + 4\beta_2\} = 1 \rightarrow \beta_{-1} + \beta_1 + 4\beta_2 = 2/3 \\
&\alpha_1 + 4 \cdot \{\beta_{-1} - \beta_1 - 8\beta_2\} = 1 \rightarrow \beta_{-1} - \beta_1 - 8\beta_2 = 0
\end{aligned}$$

In this case, these five equations lead to $\beta\,0 = 4/3$, $\beta_{-1} = \beta_1 = 1/3$, and $\beta_2 = 0$, which yields Simpson's rule. The local truncation error can be found by assuming it has the form:

$$E_t = C \cdot (h)^5 \cdot f^{(5)}(x) = C \cdot (h)^5 \cdot 120$$

And substituting $y = x^5$ into the Simpson rule expression, we obtain:

$$(x+h)^5 = (x-h)^5 + \frac{5}{3} h \cdot \left\{(x+h)^4 + 4x^4 + (x-h)^4\right\} + E_t \rightarrow E_t = -\frac{4}{3} h^5$$

And equating the two previous equations gives $C = -1/90$, as in Example 3.1.

3.3.2 *Most Used Algorithms: The Adams Families*

Among explicit MLMs, the most popular are the Adams–Bashforth family of predictors of the form:

$$y_{i+1} = y_i + h \cdot \sum_{k=0}^{p} \beta_k f_{i-k}. \tag{3.18}$$

The coefficients β_j and C for methods of order 1 through 6 are shown in Table 3.1.

Table 3.1 Coefficients of the Adams–Bashforth predictors

Order p	Scale factor	β_0	β_1	β_2	β_3	β_4	β_5	C
1	1	1						1/2
2	2	3	−1					5/12
3	12	23	−16	5				9/24
4	24	55	−59	37	−9			251/720
5	720	1901	−2774	2616	−1274	251		475/1440
6	1440	4277	−7923	9982	−7298	28,771	−475	19,087/60480

For example, for the fourth-order method $\beta_2 = 37/24$ and the local truncation error is $251 \cdot h^5 \cdot f^{(5)}(\xi)/720$. Note that the first-order predictor corresponds to the explicit Euler method, Eq. (3.7).

The corresponding correctors form the Adams–Moulton family, given by:

$$y_{i+1} = y_i + h \cdot \sum_{k=-1}^{p} \beta_k f_{i-k} \tag{3.19}$$

Which are also frequently used. Table 3.2 presents the coefficients for these methods of orders 2 to 6. The first-order method corresponds to the implicit Euler method, and the second-order method corresponds to the trapezoidal rule.

The corrector of Eq. (3.19) is implicit in y_{i+1} and must be solved by some iterative procedure, as discussed in Chap. 2. For these cases, fixed-point iteration is typically used, in which Eq. (3.19) is written in the form

$$y_{i+1}^{(j+1)} = h \cdot \beta_{-1} \cdot f\left(x_{i+1}, y_{i+1}^{(j)}\right) + B = \phi\left(y_{i+1}^{(j)}\right) \tag{3.20}$$

where B includes all remaining terms (known) on the right-hand side of Eq. (3.19) and j is the iteration counter at the integration step. According to Chap. 2, the convergence criterion for Eq. (3.20) is that the absolute value of the derivative of ϕ be less than one:

Table 3.2 Coefficients of Adams–Moulton correctors

Order p	Scale factor	β_{-1}	β_0	β_1	β_2	β_3	β_4	C
2	2	1	1					−1/12
3	12	5	8	−1				−1/24
4	24	9	19	−5	1			−19/720
5	720	251	646	−264	106	−19		−27/1440
6	1440	475	1427	−798	482	−173	27	−863/60480

$$h \cdot \left| \beta_{-1} \left(\frac{\partial f}{\partial y} \right) (x_{i+1}) \right| < 1 \tag{3.21}$$

To achieve rapid convergence and keep the number of f evaluations small, h should not exceed about 10% of the bound given by Eq. (3.21). For systems of ODE, the vector-form convergence of the Eq. (3.20) requires that h satisfies the criterion

$$h \cdot |\beta_{-1} \lambda_{\max}| < 1 \tag{3.22}$$

where $\lambda_{\text{máx}}$ is the largest (in magnitude) eigenvalue of the Jacobian of the differential-equation system. If $\lambda_{\text{máx}}$ is very large, Newton-type methods should be employed to solve the nonlinear equations, if one wishes to use reasonably large step sizes h (see Sect. 3.5).

3.3.3 Predictor–Corrector Algorithms

A comparison of the coefficient C for Adams predictor and corrector methods of the same order shows that, in general, implicit formulas are considerably more accurate than explicit formulas. To take advantage of the improved accuracy of a closed formula while simultaneously reducing the number of iterations required for convergence, an open formula of the same order is used to produce the first estimate $y_{i+1}^{(0)}$ for the closed formula. Such algorithms, which employ a predictor to obtain an initial estimate of the function y and then a corrector to generate the final numerical solution at the next step, are called predictor–corrector methods, and their implementation is shown in Table 3.3.

Local truncation error estimates are also usually evaluated. For LMM, this can be done simply and inexpensively by applying Richardson extrapolation. These local truncation error estimators can be used to adjust the step size once the user has specified the acceptable error.

Table 3.3 Predictor–corrector algorithm

Step	Action
1	Compute $y_{i+1}^{(0)}$ using a predictor, for example, Eq. (3.18)
2	Compute $f\left(x_{i+1}, y_{i+1}^{(0)}\right)$ and iteratively solve the equation corresponding to the corrector (e.g., Eq. (3.19)), using scheme (3.20) until a convergence criterion is satisfied, for example: $\left| \frac{y_{i+1}^{(j+1)} - y_{i+1}^{(j)}}{y_{i+1}^{(j+1)}} \right| < \varepsilon$, where ε is a small user-defined number. Let $y_{i+1}^{(k)}$ denote the converged value
3	Increment i and repeat from step 1

In addition, if the truncation error for the predictor does not change appreciably from one step to the next, the predicted value $y_{i+1}^{(0)}$ can be modified to provide a better initial estimate for the corrector. This does not affect the corrector, except to (usually) reduce the number of iterations required for convergence. In fact, the step size h is chosen to allow the corrector to converge in only one iteration.

Note that, in this modified predictor–corrector (PC) algorithm, the derivative f is evaluated twice per step. This is half the number required by fourth-order explicit Runge–Kutta methods. Moreover, the value $y_{i+1}^{(k)}$ obtained from convergence of the corrector can in turn be modified, yielding a four-stage-per-step algorithm, PECE: (1) predict, (2) evaluate, (3) correct, and (4) evaluate again.

In Python, the method "LSODA" within *solve_ivp* includes the PECE Adams–Bashforth–Moulton algorithm with variable method order to satisfy local error criteria, developed by Shampine and Gordon (1975). This method within *solve_ivp* is useful for solving IVPs when high accuracy is required, or when the function $f(x,y)$ is expensive to evaluate.

3.3.4 Conclusions Regarding Linear Multistep Methods

The disadvantages of linear multistep method (LMM) algorithms are:

1. They are not self-starting; initial values must be generated using a Runge–Kutta (RK) method of similar order, or an LMM of lower order with a smaller step size.
2. The integration step size cannot be changed as easily as in RK methods. Variable-step LMMs can be formulated, but they have greater computational complexity than fixed-step versions.
3. Owing to the implicit nature of corrector methods, some type of iteration is required.
4. LMMs are more complicated to program than RK methods; moreover, the relevant comparison here is that LMMs require storage of prior information (historical values y_i and f_i).

On the positive side, LMMs provide local truncation error estimates and require fewer evaluations of f per step (compared with RK methods of the same order or accuracy). This has made them more popular than RK methods, particularly for systems of differential equations with widely distributed time constants (stiff systems) that are challenging to solve numerically, as will be discussed in Sect. 3.5.

3.4 Numerical Stability

Ideally, one would like to choose an algorithm with a step-size control strategy that delivers the required accuracy at minimal computational cost. If the derivatives are complicated, then the cost will be proportional to the number of function evaluations per step and the number of steps, assuming that storage operations are relatively inexpensive.

The integration step-size adjustment is normally controlled through local truncation error estimators, as described for RK and LMM methods, if the local error dominates the discretization error at each integration step. Can such an assumption be justified?

To understand the behavior of errors, consider the global discretization error for the explicit Euler method, Eq. (3.8):

$$E_{i+1} = E_i \cdot \left\{ 1 + h \cdot \left(\frac{\partial f}{\partial y} \right)(x, \alpha) \right\} - \frac{h^2}{2} f'(\xi, y(\xi))$$

with $\alpha \in \left[y_i, |y(x_i) \right]$ and $\xi \in \left[x_i, |x_{i+1} \right]$. Clearly, if the propagated error in the first term dominates the local truncation error (second term), then any assumption about the magnitude of E_{i+1} based on the methods presented in Sects 3.2 or 3.3 would be invalid. Indeed, one can see that the propagated error could be so large that the numerical solution thus generated is meaningless.

3.4.1 *Stability Analysis for the Explicit Euler Method*

Consider the solution of the ODE:

$$y'(x) = \lambda x; \quad y(0) = 1 \tag{3.23}$$

for which the exact solution is

$$y(x) = \exp(\lambda x). \tag{3.24}$$

Substituting Eq. (3.23) into Eq. (3.8) yields:

$$E_{i+1} = E_i \cdot (1 + h\lambda) - \frac{(h\lambda)^2}{2} \cdot \exp(\lambda \xi). \tag{3.25}$$

If λ is positive ($h > 0$), then E_i is amplified by the factor $(1 + h\lambda)$, which is the method's propagation factor. This result is consistent with the original problem, since $y(x)$ in Eq. (3.24) grows without bound as x increases. However, note the difficulty that arises when λ is negative: if $(1 + h\lambda) < -1$, the error will grow exponentially while alternating sign from one step to the next, whereas the exact solution decays exponentially to zero. One would expect any generated error to decay along with the solution, which will occur only if

$$|1 + h \cdot \lambda| < 1. \tag{3.26}$$

This can also be seen by substituting Eq. (3.23) directly into Eq. (3.7), which defines the explicit Euler method:

$$\begin{aligned} y_{i+1} &= y_i + h \cdot f(x_i, y_i) = y_i + h \cdot \lambda \cdot y_i = (1 + h \cdot \lambda) \cdot y_i \\ &= (1 + h \cdot \lambda)^i \cdot y_0 = (1 + h \cdot \lambda)^i \end{aligned} \tag{3.27}$$

The numerical solution in Eq. (3.27) will be valid only to the extent that $\left(1+h\lambda\right)^i$ is a close approximation to $\exp(ih\lambda) = \exp(\lambda x)$. If λ is negative and h is adjusted so that Eq. (3.26) is satisfied, then Eq. (3.27) will decay in magnitude; if we additionally want monotonic decay, the more restrictive condition must hold:

$$h < 1/|\lambda| \tag{3.28}$$

In conclusion, for this simple model in Eq. (3.23), the Euler method has a bounded interval of values of $h\lambda$ for which λ is negative (i.e., when the solution decays) and the numerical method operates properly. On the other hand, if for a given $\lambda < 0$ one uses an excessively large integration step size h, the errors will propagate without bound and will distort the numerical solution. The method will be unstable.

Although there are several definitions of stability, here we are interested in two types: relative and absolute.

(a) Relative stability is important when the exact solution of the ODE is growing (e.g., Eq. (3.24) with $\lambda > 0$); if the method is stable, the errors introduced (in y_0 or later) should not grow faster than the exact solution (in a relative sense).
(b) In contrast, absolute stability is critical when the exact solution decays: any introduced error must decay along with the solution; otherwise, the propagated error will mask the exact solution.

3.4.2 *Stability Criteria and Regions*

A similar analysis carried out for LMM using the model problem in Eq. (3.23) leads to the p th-order difference equation (with $\alpha_{-1} = -1$):

$$\sum_{k=-1}^{p} (\alpha_k + h\lambda\beta_k) y_{i-k} = 0 \tag{3.29}$$

The characteristic equations for Eq. (3.29) generate a polynomial of degree p in $(h\lambda)$ with p roots. One of these corresponds to the correct solution (the principal root γ_1); the remaining $(p-1)$ roots are spurious and are produced by the numerical approximation. Through an example, we show how the solution of the characteristic equation is obtained and how the results are interpreted.

Example 3.3 Stability region for Milne's method

Compute the roots of the characteristic Eq. (3.29) for the Milne corrector, i.e., Simpson's rule.

Substituting the model Eq. (3.23) into Simpson's-rule expression gives:

$$y_{i+1} = y_{i-1} + \frac{h \cdot \lambda}{3}\{y_{i+1} + 4y_i + y_{i-1}\} = 0$$

$$\rightarrow y_{i+1} \cdot \left\{1 - \frac{h \cdot \lambda}{3}\right\} - \frac{4h \cdot \lambda}{3} y_i - y_{i-1} \cdot \left\{1 + \frac{h \cdot \lambda}{3}\right\} = 0$$

Assuming solutions of the form $y_i = \gamma^i$, the following quadratic equation is obtained:

$$\left\{1 - \frac{h \cdot \lambda}{3}\right\} \cdot \gamma^2 - \frac{4h \cdot \lambda}{3} \cdot \gamma - \left\{1 + \frac{h \cdot \lambda}{3}\right\} = 0$$

$$\rightarrow \gamma_1 = \frac{\frac{2h \cdot \lambda}{3} + \sqrt{1 + \frac{(h \cdot \lambda)^2}{3}}}{1 - \frac{h \cdot \lambda}{3}}; \quad \gamma_2 = \frac{\frac{2h \cdot \lambda}{3} - \sqrt{1 + \frac{(h \cdot \lambda)^2}{3}}}{1 - \frac{h \cdot \lambda}{3}}$$

Figure 3.4 shows a plot of both solutions as a function of $h\lambda$. The first root γ_1 approximates the exact solution $y = \exp(h\lambda)$, whereas the root γ_2 represents behavior that is very different from the desired solution. For $h\lambda < 0$ the second root will contribute to the numerical solution by alternating sign and increasing in absolute value at each integration step, while the exact solution decreases monotonically toward zero. In other words, Simpson's quadrature rule is not a stable formula for integrating Eq. (3.23) with $\lambda < 0$.

Returning to the general solution of Eq. (3.29) and assuming that all roots of the characteristic equation are distinct, the numerical solution will have the form (Dahlquist & Bjork, 1974):

$$y_i = \sum_{k=1}^{p} C_k (\gamma_k)^i. \tag{3.30}$$

In the limit as $h \rightarrow 0$, C_1 in Eq. (3.30) approaches $y(x_0)$, whereas the coefficients $C_2, C_3, \ldots, C_p$ tend to zero (consider the case $i = 0$ in the previous equation). For small (but finite) values of h, the coefficients $C_2, C_3, \ldots, C_p$ should be small; however, for the spurious solutions to remain small relative to the principal solution $C_1\gamma_1$, it is essential that

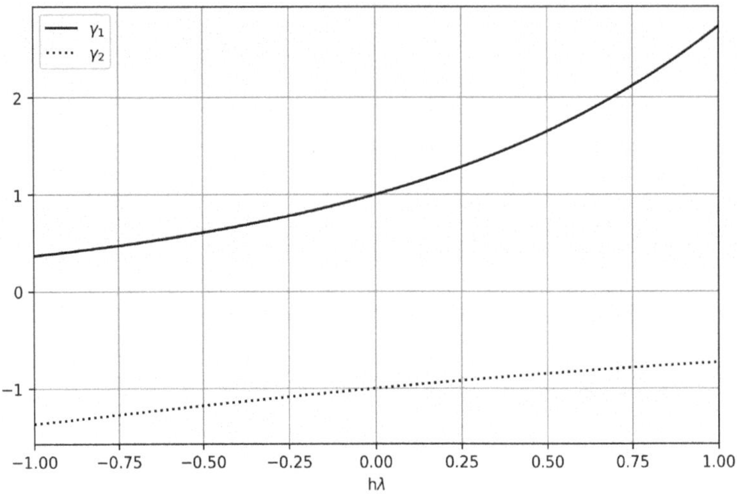

Fig. 3.4 Roots of the characteristic equation for Simpson's rule

$$|\gamma_1| > |\gamma_j|; \quad j = 2, 3, \ldots, p. \tag{3.31}$$

Since $\gamma_1 \rightarrow 1$ as $h \rightarrow 0$, , all roots of the characteristic equation (when $h = 0$) must lie on or inside the unit circle (here we assume that λ may be complex).

Therefore, a necessary condition for the convergence of a numerical method is that no root of the characteristic Eq. (3.29) with $h = 0$ lies outside the unit circle, and that the roots of magnitude 1 are simple, not multiple (Ralston & Rabinowitz, 1978). The principal root for the case $h = 0$ always has magnitude one (see the previous example, which satisfies this condition). For finite values of h, the roots may lie outside the unit circle. Two cases are relevant for stability:

(a) If the real part of λ is positive, then $|\gamma_1| > 1$, , and for the principal solution to dominate the parasitic solutions, condition (3.31) must hold. The method is relatively stable.
(b) If the real part of λ is negative, then the following condition must hold:

$$|\gamma_j| < 1; j = 1, 2, \ldots, p \tag{3.32}$$

This condition is referred to as absolute stability. Equation (3.32) simply ensures that all spurious components decay along with the principal component when the exact solution decays. Note that, in Example 3.4, γ_2 does not satisfy this condition (Fig. 3.4). Therefore, the coefficients in LMMs must be chosen so that the spurious solutions do not grow as the numerical integration proceeds. All classical LMM have been analyzed for the model Eq. (3.23); instabilities in most methods occur when the exact solution Eq. (3.24) decays while the other roots remain finite, producing a spurious solution (see Example 3.4). This can occur when $\text{Re}(\lambda) < 0$; consequently,

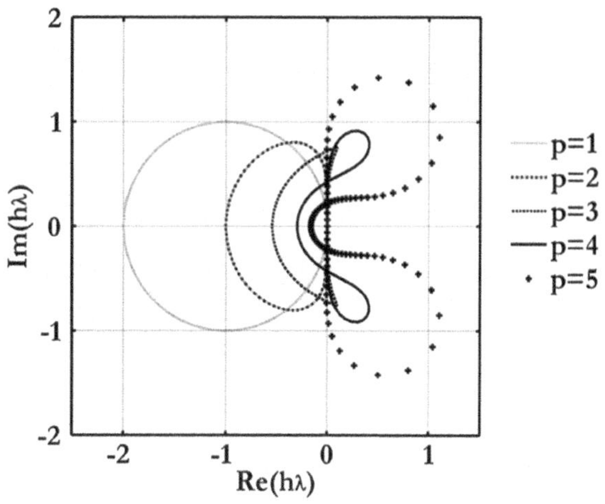

Fig. 3.5 Absolute-stability regions for the model problem (3.23) and the Adams–Bashforth methods of orders 1 through 5

stability plots for the various methods focus on the left half-plane of the complex $h\lambda$ plane.

Figure 3.5 shows the stability regions for Adams–Bashforth predictors of orders 1 through 5; the stability zone corresponds to the region inside the respective curves. Note that, for the explicit Euler method Eq. (3.7), the stability region is given by the interior of the circle defined by Eq. (3.26) in the complex plane.

As the method order increases, its accuracy increases, but the stability region shifts toward the right side of the plane. Therefore, when applying more accurate explicit methods, it is necessary to control the integration step size h so that $h\lambda$ remains within the stability region, which implies reducing the magnitude of h.

Figure 3.6 shows the absolute-stability regions for Adams–Moulton correctors of orders 1 through 5. The stability region corresponds to the exterior of each curve; for the second-order corrector (the trapezoidal rule), the stability region corresponds to the entire half-plane $\mathrm{Re}(\lambda) < 0$, as shown in the following example.

Example 3.4 Stability region for the trapezoidal rule

Consider the second order Adams corrector (the trapezoidal rule). The difference equation for the model problem Eq. (3.23) has the characteristic equation

$$\gamma \cdot (1 - h\lambda/2) - (1 + h\lambda/2) = 0 \rightarrow \gamma = \frac{1 + h\lambda/2}{1 - h\lambda/2}.$$

Therefore, the simple root is absolutely stable on the left side of the complex $h\lambda$ plane. A method with this property is called A-stable. It is important to note that, in the limit as $h\lambda \rightarrow -\infty$, $\gamma \rightarrow -1$. Although the trapezoidal rule remains stable for negative $\mathrm{Re}(h\lambda)$, the stability limit is approached as $\mathrm{Re}(h\lambda)$ increases in magnitude; the numerical solution of Eq. (3.23) will not decay to zero when λ is large and large steps are taken, but will instead oscillate, changing sign at each step and thus approaching zero in an alternating manner.

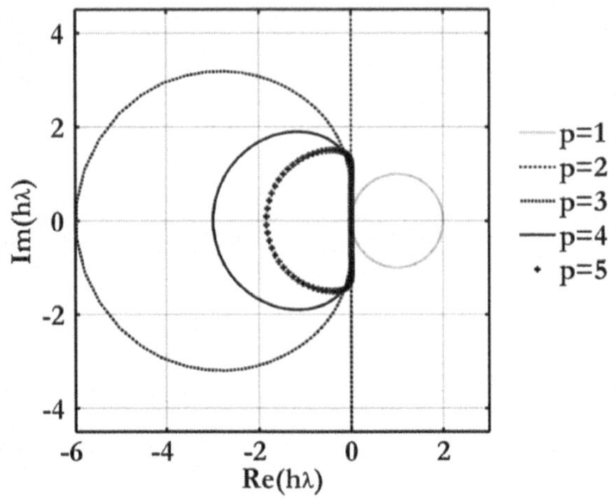

Fig. 3.6 Absolute-stability regions for the model problem (3.23) and the Adams–Moulton correctors of orders 1 through 5

Example 3.5 Stability region for the implicit Euler method

Another stable low-order method is the backward (or implicit) Euler method, which is the first-order Adams–Moulton corrector:

$$y_{i+1} = y_i + h \cdot f(x_{i+1}, y_{i+1})$$

This algorithm must be solved iteratively, and the corresponding characteristic equation has the single root

$$\gamma = \frac{1}{1 - h \cdot \lambda}.$$

This method is A-stable (Fig. 3.6). In addition, $\gamma \to 0$ as $\mathrm{Re}(h\lambda) \to -\infty$, so that the numerical solution of Eq. (3.23), when $\mathrm{Re}(h\lambda)$ is large and negative, decays to zero (even when h is large), just as the exact solution does. A method with this characteristic is called strongly A-stable (sometimes referred to as L-stable).

Runge–Kutta methods can also be analyzed for stability. Since these methods are one-step methods, the associated difference equation is always first order. Because Runge–Kutta methods have only one root, there are no spurious solutions, and it is not meaningful to discuss relative stability. Nevertheless, all explicit RK methods have absolute stability limits, since it is not possible to approximate an exponential function with a finite polynomial. The limits for some of them are shown in Fig. 3.7. Note that as the method order increases, the stability region increases in size and includes the regions of lower-order methods. This is a consequence of the increasing accuracy of the Taylor series approximation underlying these methods.

To conclude this section, note that the entire analysis has been carried out for a simple linear ODE such as Eq. (3.23). Most real problems in Eq. (3.3) are neither

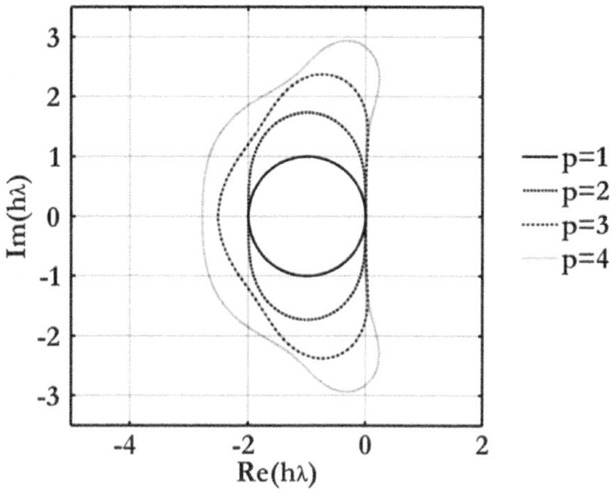

Fig. 3.7 Absolute-stability regions for the model problem (3.23) and explicit Runge–Kutta methods of orders 1 through 4

linear nor simple, so one might question these analyses. However, as shown by Hildebrand (1987), for most nonlinear ODE the nature of the error-propagation process is like what occurs in the (local) linear approximation of the ODE, at least over a small interval around x_i. Therefore, one would expect the chosen method to behave as it would for the model Eq. (3.23) if λ is replaced by $(\partial f/\partial y)$ evaluated at x_i. In practice, computational routines estimate the magnitude of $(\partial f/\partial y)$ at each integration step to decide how to change the step size h in the next step, or to switch the integration method to a more appropriate one; for further details, see Petzold (1983).

3.5 Differential Equations with Widely Separated Time Scales (Stiff Systems)

Some ODE systems cannot be solved with traditional algorithms (RK, Adams–Bashforth) because the absolute-stability regions force the use of extremely small values of h or, equivalently, an excessively large number of steps to integrate Eq. (3.3) over a finite interval. This is easy to diagnose, since the numerical solution progresses very slow when commonly used functions are applied; when this occurs, it is a symptom of stiff ODE problems.

Such systems of ODE are usually characterized by solutions consisting of components that vary slowly with the independent variable (usually time) and others that vary rapidly. Examples include circuit dynamics, chemical reactions, phase equilibrium in open systems, atmospheric chemistry and distillation, where multiple phenomena occur on different time scales. The components that vary rapidly with time are transients that decay quickly to a constant value, whereas the more slowly varying components eventually tend toward a steady-state solution. ODE systems with these characteristics are referred to as stiff, because when the solution is plotted as a function of time, the faster-varying components appear as sharp rises or drops ("stiff" behavior), followed by much smoother curves. Although the transient components may be of physical importance only over a small portion of the integration interval, they can restrict, over the entire interval, the maximum value of h that can be used, due to the stability condition of the integration method. Example 3.6 illustrates the case for a single differential equation.

Example 3.6 Stiff differential equation

The equation

$$y'(x) = -10^5 \cdot y + 10^5 \cdot \exp(-x) + \exp(-x); \quad y(x_0) = 0$$

has the analytical solution

$$y(x) = \exp(-x) - \exp(10^5 \cdot x)$$

where $\exp\left(-10^5 x\right)$ is a rapidly decaying component superimposed on a more slowly varying component, $\exp(-x)$. For $x > 1.4 \times 10^{-4}$, the faster component is 10^{-6} times the value of the slow component and does not influence the exact solution $y(x)$, as shown in Fig. 3.8.

Suppose we wished to use the Euler method in Eq. (3.7) to solve the previous differential equation; the propagation factor in Eq. (3.8) would apply, and the stability criterion would restrict h such that

$$\left|1 - 10^5 \cdot h\right| < 1 \rightarrow h < 2 \times 10^{-5}$$

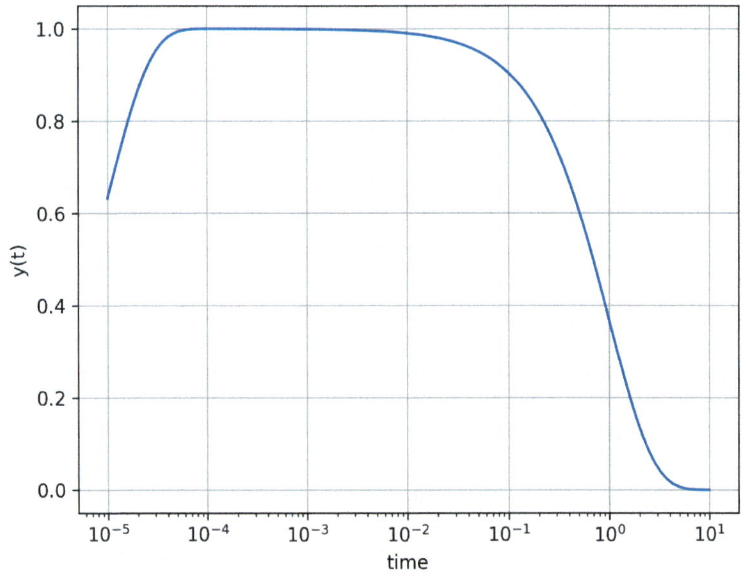

Fig. 3.8 Solution of the differential equation in Example 3.6

Thus, to resolve the initial transient component accurately, one would have to use a small value of h(much less than 10^{-5}); however, once this faster component becomes negligible, it would seem reasonable to use larger values of h to compute the part of the solution where e^{-x} dominates.

The preceding criterion indicates that this is not the case. If one attempts to use values of h larger than 2×10^{-5} after the initial transient has decayed, unstable error propagation caused by the fast component will mask the numerical solution, producing erroneous results. For example, if $h = 10^{-4}$, the fast component is multiplied by a factor of -9 at each step, growing without bound.

This is a typical feature of stiff systems when one attempts to solve them with inappropriate methods: classical explicit solvers (RK23, RK45) become excessively slow because they reduce the integration step size to satisfy the tolerance criterium for the local error.

In contrast, the implicit Euler method guarantees that, regardless of the value of h, the method will produce a stable solution that eventually approaches $\exp(-x)$. Of course, if we also require accuracy over the first transient interval, h should be small. Once it has decayed, h can be increased again with the assurance that the rapid transient component will continue to decay; even if it is not computed accurately, it will be so small in absolute value that it will not affect the numerical solution.

Although the stiff condition can occur in the solution of a scalar ODE as in Example 3.6, it typically arises in systems of ODE whose solutions contain components that evolve on widely separated time scales. The simplest case is the decoupled system of n ODE analogous to the model problem in Eq. (3.23), which can be written in matrix notation as:

$$y' = f(x, y) = A \cdot y \tag{3.33}$$

where $A = \text{diag}(\lambda_1, \ldots, \lambda_n)$. This matrix can be considered as the Jacobian for the system of equations in Eq. (3.33). Since J is in diagonal form, $\{a_{jj}=\lambda_j\}$ are the eigenvalues of the Jacobian, and they are also called the eigenvalues of the ODE system.

The solution to Eq. (3.33) is

$$y_j = y_j(0) \cdot \exp(\lambda_j \cdot x) \tag{3.34}$$

If Eq. (3.33) is solved using the explicit Euler method, the algorithm is analogous to Eq. (3.7):

$$y_{n+1} = (I_n + h \cdot A) \cdot y_i \tag{3.35}$$

Here, I_n is the $n \times n$ identity matrix. The absolute stability criterion analogous to Eq. (3.24) is

$$|1 + h \cdot \lambda_i| < 1; \quad j = 1, 2, 3, \ldots, n \tag{3.36}$$

Note that the above equation requires that step h must satisfy (if all $\mathrm{Re}(\lambda_j) < 0$)

$$0 < h < \min\{2/|\lambda_i|\}; \quad j = 1, 2, 3, \ldots, n \tag{3.37}$$

Therefore, the step h is limited by the eigenvalue of largest magnitude for the ODE system.

For the implicit Euler method, the update is

$$y_{i+1} = (I + h \cdot A)^{-1} \cdot y_i \tag{3.38}$$

As in the scalar case, this method is A-stable, since h is not constrained by stability restrictions.

3.5.1 Appropriate Methods for Stiff ODE

If one wishes to solve a stiff problem with reasonable computational cost, it is essential that the chosen algorithm be A-stable (or strongly A-stable) or a good approximation to this class of stability. Therefore, in this case, classical methods such as explicit RK or the PECE implementation of the Adams methods simply cannot be used due to their small region of absolute stability in the left-half plane (see Figs. 3.5 and 3.7).

Semi-implicit RK methods have appropriate stability (and are highly accurate) for solving stiff problems, but they have not been used intensively. This is due to the difficulty in estimating the local truncation error of the method. The references cited in Sect. 3.2.3 provide details on RK methods that have wide stability regions but are semi- or fully implicit.

Gear (1971) was the first to develop numerical methods suitable for stiff problems. He defines that any numerical method which is stable for solving stiff problems for the model Eq. (3.23) must satisfy the conditions shown in Fig. 3.9.

Thus, when $h\lambda$ lies in region II, the method must be both stable and accurate; in region I, accurate; and in region III, stable. In this way, a solution component with a large negative λ(a rapidly decaying transient) can be computed precisely using a sufficiently small h; once that component has become small in magnitude, it will remain stable even if h is increased.

3.5.2 Implementation of Algorithms for Stiff ODE

Gear (1971) has developed several high-order LMM that have near A-stable behavior; these methods are now available in several software packages, including Python. Gear's p-th order backward differentiation formula has the form:

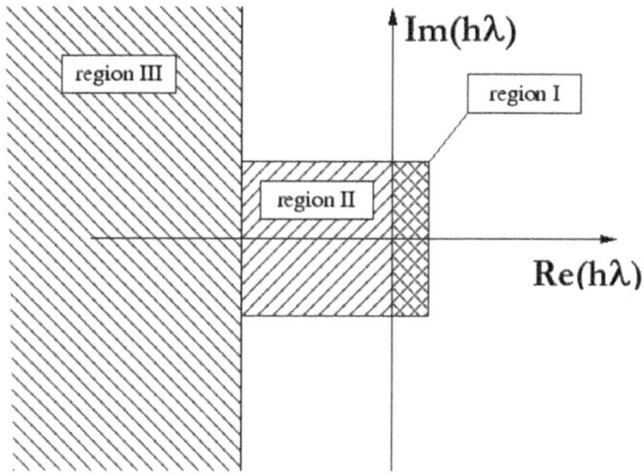

Fig. 3.9 Stability requirements for stiff ODE

$$y_{i+1} = \sum_{k=0}^{p-1} \alpha_k \cdot y_{i-k} + h \cdot \beta_{-1} \cdot f_{i+1}. \tag{3.39}$$

The name of the method arises from the fact that the above equation can be interpreted as an estimate of the derivative $y'(x_{i+1})$ in terms of present and past values. The method is implicit in y_{i+1}, but it uses only one evaluation of f; these backward-difference formulas are, therefore, LMM correctors. The lowest order one is the implicit Euler method. A summary of formulas is given below in Table 3.4.

In most cases, of course, the ODE will not only be coupled, but they will also be nonlinear. Consequently, solving the algebraic equations generated by an implicit algorithm poses difficulties, since the equations are nonlinear. An iterative algorithm is required.

A typical case is the implicit Euler method, which requires solving for the j-th component of the ODE system.

$$y_{j,i+1} = y_{j,i} + h \cdot f_j(\vec{x}_{i+1}, \vec{y}_{i+1}); \quad j = 1, 2, \ldots, n \tag{3.40}$$

One might try to treat Eq. (3.40) as a typical corrector and write the fixed-point iteration:

Table 3.4 Coefficients of Gear's correctors

Orden p	α_0	α_1	α_2	α_3	α_4	α_5	β_{-1}
1	1						1
2	4/3	−1/3					2/3
3	18/11	−9/11	2/11				6/11
4	48/25	−36/25	16/25	−3/25			12/25
5	300/137	−300/137	200/137	−75/137	12/137		12/137
6	360/147	−450/147	400/147	−225/147	72/147	−10/147	60/47

$$y_{j,i+1}^{(m+1)} = y_{j,i} + h \cdot f_j\left(x_{i+1}, y_{i+1}^{(m)}\right) = \phi\left(y_{i+1}^{(m)}\right) \tag{3.41}$$

where m is the iteration counter. Unfortunately, the restrictions on the step size h required for the convergence of Eq. (3.41) are very stringent when compared with the absolute stability criteria, which allow unbounded values of h.

In other words, Eq. (3.41) cannot be used successfully for stiff problems. The implicit corrector equations are solved using Newton–Raphson-type methods. In the case of the classical Newton–Raphson method, one must solve the linear equations

$$\boldsymbol{J}^*\left(\boldsymbol{x}_{i+1}, \boldsymbol{y}_{i+1}^{(m)}\right) \cdot \boldsymbol{\delta}^{(m)} = -\boldsymbol{F}\left(\boldsymbol{x}_{i+1}, \boldsymbol{y}_{i+1}^{(m)}\right) \tag{3.42}$$

where

$$\boldsymbol{F}_j(\boldsymbol{x}_{i+1}, \boldsymbol{y}_{i+1}) = \boldsymbol{y}_{j,i+1} - \boldsymbol{y}_{j,i} - h \cdot \boldsymbol{f}(\boldsymbol{x}_{i+1}, \boldsymbol{y}_{i+1}). \tag{3.43}$$

And J^* is the Jacobian of the system of equations $F(x_{i+1}, y_{i+1}) = 0$. Iteration (3.42) is carried out until $\delta^{(m)}$ becomes sufficiently small in norm. Fortunately, the Jacobian does not need to be updated frequently; it is sufficient to use an approximate Jacobian. This can substantially reduce the computational time, since evaluating/inverting J^* is expensive for large systems. Note that the Jacobian J^* of the implicit method (for Gear's backward differentiation formulas) is related to the Jacobian J of the original ODE system by:

$$\boldsymbol{J}^* = \boldsymbol{I} - \boldsymbol{\beta}_{-1} \cdot \boldsymbol{J} \tag{3.44}$$

Figure 3.10 shows the stability regions for Gear's backward differentiation methods of orders 1 through 6; the stability regions are those lying outside the corresponding curves.

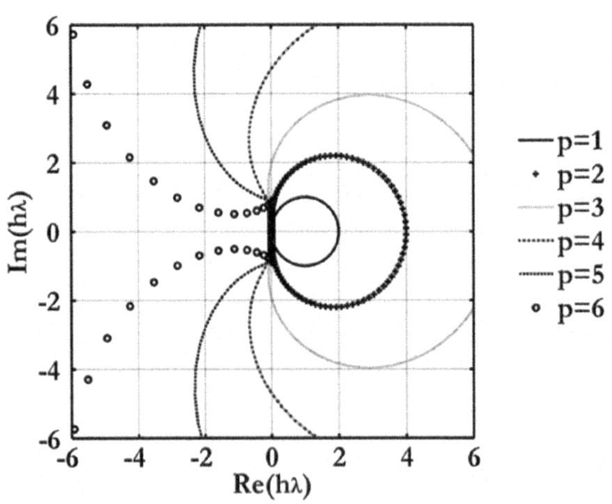

Fig. 3.10 Absolute stability regions for model problem (3.23) and backward differentiation integration methods of orders 1 to 6

Python provides three basic methods for stiff problems: BDF, Radau, and LSODA within *solve_ivp*. The first is appropriate when the system is stiff, the second one when the system is extremely stiff and the third one is adequate when we do not know how stiff the system might be.

3.6 Selection of a Numerical Integration Method

Users of general-purpose initial value problem (IVP) software have a wide variety of options when selecting a method, an error-control strategy, and so forth. Some of the questions that must be addressed are:

1. Is it necessary to solve stiff cases or not?
2. What class of methods will be used?
3. Will the method order be fixed or variable?
4. Will the method be changed during the integration?
5. How are the starting values determined?
6. How will h be adjusted to keep errors within the user's specifications?

With respect to the last question, we must distinguish between ODE that are not stiff and those that are:

(a) If the system is not stiff, several considerations affect the integration step. First, a step must be chosen so that the formula is accurate at the next step (the local truncation error is less than a small constant ε). If the code uses functional iteration to solve the corrector equation, the step must be sufficiently small to guarantee convergence. Finally, the step must be sufficiently small for the method to be stable.

(b) If the system is stiff, the integration step is chosen so that the formula is accurate. It is usually assumed that the convergence of Newton's method does not impose additional constraints on the step size.

In both cases, the option remains to increase the method order to achieve higher accuracy without further reducing the integration step size. One of the first general-purpose codes was EPISODE, developed by Hindmarsh et al. (1976). It uses variable step size and variable order (including a stiff/non-stiff option). It is also possible to write codes that automatically switch methods depending on the local character of the system (stiff or not), such as the LSODE codes of Hindmarsh and Petzold (1995). The latter also allows the incorporation of nonlinear equations (such as thermodynamic equilibrium constraints) that must be satisfied along the integration trajectory; in this case, the software is referred to as DASSL (Petzold, 1982). For differential equations with algebraic constraints, Python interface Assimulo includes routine IDA which handles nonlinear constraints along the integration trajectory.

Tables 3.5 and 3.6 summarize the criteria presented in this chapter and serve as a quick reference for the user.

Table 3.5 Selection of a numerical integration method

Features of the system $y' = f(x,y)$	Multistep methods		One step methods		
	PECE	Gear	RK explicit	RK-DI[§]	Rosenbrock
Small: $n \leq 10$	✓	✓	✓	✓	✓
Large: $n > 10$	✓	✓	✓	✓	
(∂f/∂y) "small"	✓		✓		
(∂f/∂y) "large"		✓		✓	✓
Non oscillatory	✓	✓	✓	✓	✓
Very oscillatory				✓	✓
Stable*	✓	✓	✓	✓	✓
Unstable*			✓	✓	
$f(x,y)$ expensive	✓	✓			
$f(x,y)$ inexpensive	✓	✓	✓	✓	✓
Discontinuous in t		✓	✓	✓	✓
A lot of intermediate information	✓	✓			
High precision	✓	✓			

*From the mathematical point of view
§Stands for diagonally implicit Runge–Kutta method

Table 3.6 Characteristics of numerical integration methods for IVPs

Method/class	Accuracy	Self-starting?	Stability on real axis
Euler/explicit	$O(h)$	Yes	(−2,0)
RK 2° order/explicit	$O(h^2)$	Yes	(−2,0)
RK 4° order/explicit	$O(h^2)$	Yes	(−2.78,0)
Adams-B./explicit	$O(h^4)$ $(p = 4)$	No	(−0.3, 0)
Adams-B./exp	$O(h^8)$ $(p = 8)$	No	(−0.03, 0)
Euler/implicit	$O(h)$	Yes	$(-\infty,0)$ U $(2, \infty)$
Trapezoidal/implicit	$O(h^2)$	Yes	$(-\infty, 0)$
Adams-M./implicit	$O(h^4)$ $(p = 3)$	No	(−3, 0)
Adams-M/implicit	$O(h^8)$ $(p = 7)$	No	(−0.5, 0)
RK/D. implicit	$O(h^3)$	Yes	$(-\infty,0)$ U $(8.2, \infty)$
Rosembrock/implicit	$O(h^3)$	Yes	$(-\infty,0)$ U (a, ∞)
Gear/implicit	$O(h) - O(h^6)$	No	$(-\infty,0)$ U (b, ∞)
Adams/PECE	$O(h^4)$	No	(−1.25,0)
Adams/P(EC) 2	$O(h^4)$	No	(−0.9,0)
Adams/PECE	$O(h^8)$	No	(−0.4,0)

3.7 Implementation of Numerical Integrators in Python

In Python, *solve_ivp* is the general-purpose routine for the numerical integration of systems of ordinary differential equations of the form:

$$y' = \mathrm{d}y/\mathrm{d}t = f(t, y);\ y(t = 0) = y_0.$$

The syntax is sol = solve_idp(fun, t_span, y_0, method, t_eval, rtol, atol) where:

- fun: name of the function that evaluates the differential equation system $y' = f(t, y)$. It must have the syntax: *yp* = fun(*t*,*y*), where *yp* is defined by the user and contains all expressions required to evaluate the derivatives $\mathrm{d}y/\mathrm{d}t$ (for each component of y).
- $t_{\text{span}} = [t_0, t_f]$: initial and final times for the numerical integration. If a time vector is provided in the *t*_eval argument, the solution will be returned to those user-specified points.
- y_0 = initial value of y at $t = t_0$(a vector).
- method: includes RK23 and RK45 for non-stiff systems, BDF and Radau for moderate and highly stiff systems, respectively, and LSODA for choosing variable-order methods such as Adams-type integrators.
- rtol, atol: relative and absolute tolerances for local errors at each integration step, respectively.

At the end of the integration, one obtains an object with the following structure:

- sol.t: column vector containing the discrete integration times; these are selected automatically to satisfy the default local error criteria built into the integrator. However, if the t_eval argument was input, sol.t contains the same times as *t*_eval.
- sol.y: solution of the system; it is a matrix with N columns (one per dependent variable in y) and M rows, corresponding to the discrete times at which the integrator returns the numerical solution.
- sol.status: a flag indicating the success of the integrator, a value '0' means a successful integration.
- sol.message: additional comments brought up by the routine.

Example 3.7 Integration of a scalar ODE using solve_idp

The following script provides an example of usage of solve_idp.

```
import numpy as np
import matplotlib.pyplot as plt
from scipy.integrate import solve_ivp

def f(t, y):
    return -2*y

sol = solve_ivp(f, (0, 5), [1], t_eval=np.linspace(0,5,100))

plt.plot(sol.t, sol.y[0])
plt.xlabel('t')
plt.ylabel('y(t)')
plt.grid()
plt.show()
```

3.8 Parameter Optimization in Dynamic Models

In this Section, we apply numerical ODE integration techniques to a type of problem frequently encountered in process engineering.

Consider the following example of a dynamic microbial growth model, described by the following system of ordinary differential equations, where $y = [X, S, O_2]^T$ is the state vector that describes biomass, substrate, and dissolved oxygen in the culture medium, respectively:

$$\begin{aligned}
&\frac{dX}{dt} = \mu X; \quad \mu = \mu_{\max}\frac{S}{K_S + S} \\
&\frac{dS}{dt} = -\frac{\mu X}{Y_{\frac{X}{S}}} - mX \quad \text{if } S > 0, \ \text{else } \frac{dS}{dt} = 0 \\
&\frac{dO_2}{dt} = K_{\mathrm{LA}}\left(O_2 - O_2^*\right) - \frac{\mu X}{Y_{\frac{X}{O2}}} - m_{O_2}X \quad \text{if } O_2 > 0, \ \text{else } \frac{dO_2}{dt} = 0
\end{aligned} \tag{3.45}$$

Note that the elements $p_1 = \mu_{\max}$, $p_2 = K_S$, $p_3 = Y_{X/S}$, $p_4 = m$, etc., are parameters of the dynamic microbial growth model. The term "parameter" is typically used to denote constants that are not derived from the fundamental conservation equations of mass, momentum, or energy, but instead arise from additional relations, such as the activation energy in the Arrhenius law for kinetic constants, the saturation constant in Monod kinetics, the overall mass-transfer coefficient between air and the liquid medium, and so forth. Indeed, these parameters are required to complete the mass, momentum, or energy balance equations so that they can be solved numerically.

Then, we can write the equations of the previous dynamic model in the following form:

$$\frac{dy_i}{dt} = f\left(t, \{y_1, \ldots, y_n\}, \{p_1, \ldots p_p\}\right); \quad i = 1, 2, \ldots, n \tag{3.46}$$

In general, the parameters $\{p_k\}$ are known; however, it often occurs that some of them are not known accurately for a particular application of a process model and must be estimated from additional information, such as experimental values of $y_i(t)$ measured at a discrete set of experimental times $\{t_1, t_2, t_3, \ldots, t_m\}$.

For example, in the case of filamentous fungi growth, the value $p_3 = Y_{X/S} = 2/3$ is common to most strains studied, whereas the mass-transfer coefficient K_La depends on the agitation and geometry of the fermenter (i.e., on the process operating conditions), and so forth. In turn, it is common for transport parameters such as K_La to be obtained from correlations of experimental data, which themselves exhibit some degree of uncertainty around a mean value.

In summary, the parameters $\{p_k\}$ are uncertain; therefore, dynamic models typically undergo a calibration stage in which the parameters are estimated by requiring that the simulated model match the experimental measurements from the process as closely as possible. Thus, the estimation of the optimal parameters $\{p_k\}^*$ can be formulated as the minimization of the following objective function (m is the number of measurement points and n is the number of components of the vector $y(t)$):

$$\min F(p_1, p_2, \ldots, p_p) = \sum_{j=1}^{m} \sum_{i=1}^{n} \left\{ (y_i(t_j))^{\mathrm{OBS}} - (y_i(t_j, \{y_k\}, \{p_k\}))^{\mathrm{MOD}} \right\}^2 \tag{3.47}$$

where the dependence of the objective function on the parameters $\{p_k\}$ is implicit in the model outputs, i.e., in the simulated values of $y_i(t)$ evaluated at the different times $\{t_1, t_2, t_3, \ldots, t_m\}$.

3.8.1 Python Implementation

Figure 3.11 illustrates the Python implementation of a search algorithm for the optimal parameters.

As an example of optimization routine available in Python, we have already mentioned *minimize*, which handles both unconstrained and constrained minimizations. This function performs a search over the parameter vector $\{p_k\}$ that minimizes the scalar function in Eq. (3.47) defined by the user in the function Integrate.

Next, an example of the implementation of the above scheme is presented.

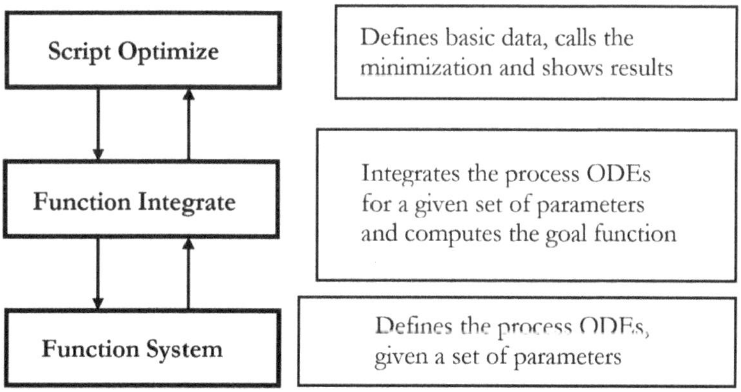

Fig. 3.11 Schematic of a parameter optimization algorithm for dynamic models

Example 3.8 Estimation of chemical kinetics parameters

For the irreversible kinetic scheme given by: $y_1 \overset{k1}{\rightarrow} y_2 \overset{k2}{\rightarrow} y_3$, and mathematically described by the system of equations:

$$\frac{\mathrm{d}y_1}{\mathrm{d}t} = -k_1 \cdot y_1; \quad \frac{\mathrm{d}y_2}{\mathrm{d}t} = k_1 \cdot y_1 - k_2 \cdot y_2; \quad \frac{\mathrm{d}y_3}{\mathrm{d}t} = k_2 \cdot y_2$$

The following data are available from two independent experiments:

t (min)	10	20	40	80	160	320
$y_2(t)$, Exp. 1	0.13	0.24	0.37	0.45	0.42	0.21
$y_2(t)$, Exp. 2	0.12	0.23	0.35	0.44	0.41	0.20

which were obtained under conditions such that $y_0 = [1\ 0\ 0]^{\mathrm{T}}$.

Determine the optimal parameters k_1 and k_2 that explain these experimental data, following the previously proposed scheme.

Figure 3.12 shows the result of the parameter estimation, where it was found that $k_1 = 0.0131\ \mathrm{min}^{-1}$ and $k_2 = 0.0073\ \mathrm{min}^{-1}$. The fit obtained by the model follows the trend of the data in a least-squares sense.

The Python script used to solve this example is presented below. The problem was assumed to be non-stiff, and the explicit integrator RK45 was used. Notice how the parameters have been input into the differential equations of the system as additional inputs. Likewise, in the call to solve_ivp, the parameters are inputs using the args option.

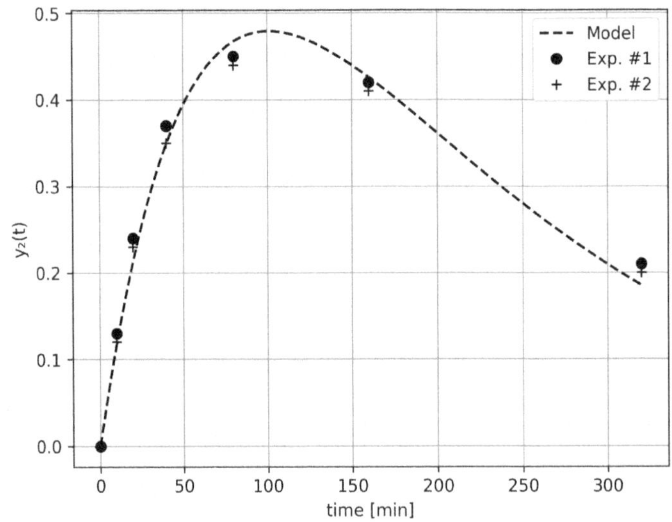

Fig. 3.12 Result of the parameter estimation for Example 3.8

```
import numpy as np
import matplotlib.pyplot as plt
from scipy.integrate import solve_ivp
from scipy.optimize import minimize

# -------------------------------------------------
# Macro to estimate kinetic constants
# -------------------------------------------------

# Experimental data
texp = np.array([10, 20, 40, 80, 160, 320])
yexp1 = np.array([0.13, 0.24, 0.37, 0.45, 0.42, 0.21])
yexp2 = np.array([0.12, 0.23, 0.35, 0.44, 0.41, 0.20])

# Initial conditions
Y0 = np.array([1.0, 0.0, 0.0])

# Initial guess for parameters
x0 = np.array([0.01, 0.005])

# -------------------------------------------------
# Function that defines the kinetic model
# -------------------------------------------------
def System(t, y, k1, k2):
    """
    Function that describes the kinetic model of the system
    """
    dy1 = -k1 * y[0]
    dy2 = k1 * y[0] - k2 * y[1]
    dy3 = k2 * y[1]
    return [dy1, dy2, dy3]

# -------------------------------------------------
# Objective function for parameter estimation
# -------------------------------------------------
def Integrate(x):
    """
    Function that computes the objective function for parameter fitting;
    in this case there are two experimental curves
    """
    k1, k2 = x

    # Time points including t = 0
    t_eval = np.concatenate(([0], texp))

    # Integrate the kinetic system
    sol = solve_ivp(
        System,
        (0, texp[-1]),
        Y0,
        t_eval=t_eval,
        args=(k1, k2),
        method='RK45'
    )

    # Model prediction for y2(t)
    y_model = sol.y[1]

    # Differences between model and experiments
    delta1 = y_model - np.concatenate(([Y0[1]], yexp1))
    delta2 = y_model - np.concatenate(([Y0[1]], yexp2))

    # Sum of squared errors
    return np.sum(delta1**2 + delta2**2)

# -------------------------------------------------
# Parameter optimization
# -------------------------------------------------
result = minimize(Integrate, x0, method='BFGS')
k1_opt, k2_opt = result.x
```

```
print(f'Estimated k1 = {k1_opt:.5f}')
print(f'Estimated k2 = {k2_opt:.5f}')

# -----------------------------------------------
# Final simulation using optimal parameters
# -----------------------------------------------
t_sim = np.arange(0, 321, 4)

solution = solve_ivp(
    System,
    (0, 320),
    Y0,
    t_eval=t_sim,
    args=(k1_opt, k2_opt),
    method='RK45'
)

# -----------------------------------------------
# Plot results
# -----------------------------------------------
plt.plot(solution.t, solution.y[1], 'k--', label='Model')
plt.plot([0, *texp], [0, *yexp1], 'ko', label='Exp. #1')
plt.plot([0, *texp], [0, *yexp2], '+k', label='Exp. #2')

plt.xlabel('time [min]')
plt.ylabel('y₂(t)')
plt.legend()
plt.grid(True)
plt.show()
```

3.9 Problems

(1) Consider the Ordinary Differential Equation:

$$\frac{\mathrm{d}y}{\mathrm{d}t} = f(t, y) = 5\left(y - t^2\right)$$
$$y(0) = 0.08$$

(a) Provide formulas for computing the numerical approximation y_{i+1} when the explicit Euler method is applied with a constant integration step h.
(b) Repeat the same, but now for the implicit Euler method.
(c) When solving the equation with $h = 0.01$, it was found that the explicit Euler method yields a negative solution that decreases without bound for $t > 1.4$; in contrast, for the implicit method, the solution increases without bound and remains always positive. Explain why these results occur.

(2) Consider the ODE system:

$$\frac{\mathrm{d}\vec{y}}{\mathrm{d}t} = \begin{bmatrix} 42.2 & 50.1 & -42.1 \\ -66.1 & -58 & 58.1 \\ 26.1 & 42.1 & -34 \end{bmatrix} \begin{bmatrix} y_1 \\ y_2 \\ y_3 \end{bmatrix};$$

$$\vec{y}(t=\pi/8)=\begin{bmatrix}\exp(-50\frac{\pi}{8})\\ \exp(\frac{\pi}{80})-\exp(-50\frac{\pi}{8})\\ -\exp(\frac{\pi}{80})+\exp(-50\frac{\pi}{8})\end{bmatrix}$$

which is to be integrated from $t=\frac{\pi}{8}$ to $t=\pi/2$, using $h=\Delta t=\pi/64$.

(a) Write a script to compute the solution at $t=\frac{\pi}{2}$ using the implicit Euler method.
(b) Repeat part (a) but now using the trapezoidal method.
(c) Repeat part (a), but for the fourth-order Runge–Kutta method.
(d) Plot the numerically obtained values of $y_1(t)$ for the three methods tested. Also include in the plot the exact solution given by:

$$y_1(t)=sen(8t)*\exp(t/10)+\exp(-50t)$$

(e) Plot the differences between each numerical integration in parts (a), (b), and (c) and the exact solution.
(f) Indicate what performance differences you observe among the three methods. What is the reason for this difference? Which method would you prefer, and why?

(3) Heun (1900) proposed the following numerical integration method for an ordinary differential equation:

$$y_n^*=y_n+hf(t_n,y_n);\; y_{n+1}=y_n+\frac{1}{2}\left[f(t_n,y_n)+f\left(t_{n+1},y_n^*\right)\right];$$

Write a function that integrates an ordinary differential equation using Heun's method. Plot the (absolute) global error at $t=1$, as a function of the integration step h, when the Heun, explicit Euler, and implicit Euler methods are used to integrate the differential equation $\mathrm{d}y/\mathrm{d}t=-y$, $y(0)=1$. Perform your calculations for $h=0.1, 0.25, 0.05, 0.01, 0.005$, and 0.0025. Indicate which method is the most accurate and which is the least accurate for this example.

(4) The pyrolysis of ethane (C_2H_6) is carried out to produce ethylene and hydrogen via the following endothermic gas-phase reaction: $C_2H_6 \rightarrow C_2H_4 + H_2$. A tubular reactor is used, to which heat is supplied through its wall at a rate of 5000 BTU/(h ft^2). Assuming plug flow of the gases and ideal-gas behavior, the equations describing the spatial evolution of the conversion x and temperature T along the reactor can be written as:

$$\frac{\mathrm{d}x}{\mathrm{d}V}=\frac{kP}{F_0RT}\left(\frac{1-x}{1+x}\right);\quad x(0)=0;$$

$$\frac{\mathrm{d}T}{\mathrm{d}V}=\frac{\left\{\frac{\pi D_i q}{A\cdot F_0}+\frac{\mathrm{d}x}{\mathrm{d}V}(-\Delta H_R)\right\}}{\{(1-x)C_{P1}+x(C_{P2}+C_{P3})\}};\quad T(0)=920\text{ K}$$

$$k = 2.1 \times 10^{20} e^{-\frac{55000}{T}}$$

$$\Delta H_R = 45000 + 16T - 0.006T^2 + 1.3 \times 10^{-6} T^3 (\text{BTU/lbmol})$$

Additional data:

$F_0 = 100$ (lbmol/h); $A = 0.09$ ft^2; $P = 30$ psia; $R = 10.73$ (psia ft^3/lbmol K); $q = 5000$ BTU/(h ft^2); $D_i = 0.225$ ft; $C_{P1} = 22$ BTU/(lbmol K); $C_{P2} = 15$ BTU/(lbmol K); $C_{P3} = 7$ BTU/(lbmol K);

(a) Write the macros required to compute the conversion and temperature profiles and determine the reactor volume required to achieve a 10% conversion at the reactor outlet.
(b) Plot the concentration profiles along the reactor for different values of the heat input rate q, between 3000 and 12,000 BTU/(h ft^2).

(5) Consider the following system of ODE corresponding to a kinetic-type problem, with different time constants for each component of the vector y:

$$\frac{dy_1}{dt} = -0.04 \cdot y_1 + 10^4 \cdot y_2 \cdot y_3$$
$$\frac{dy_2}{dt} = 0.04 \cdot y_1 - 10^4 \cdot y_2 \cdot y_3 - 3 \times 10^7 (y_2)^2$$
$$\frac{dy_3}{dt} = 3 \times 10^7 (y_2)^2$$

The initial condition for this system is: $\mathbf{y}(0) = [1\ 0\ 0]^T$.

(a) Integrate this system of equations from $t = 0$ to $t = 0.1$ using *solve_idp* with methods RK23 or RK45. Are reasonable results obtained? Does integration take a long time?
(b) Now solve the previous system using the methods BDF or LSODA. Comment on the performance of both routines compared with RK23 and RK45.
(c) Plot the results of the above numerical integrations of the system of equations, using axis scales appropriate for representing the dynamics of this system.

(6) Consider the equation:

$$y'(x) = f(x, y) = -10^5 y + 10^5 e^{-x} + e^{-x}; \quad y(0) = 0$$

which has the analytical solution (always positive):

$$y(x) = e^{-x} - e^{-100000x}$$

(a) Try to integrate this system of equations from $x = 0$ to $x = 1$ using RK23 or RK45. Are the results reasonable? Does integration take a long time?
(b) Now try to solve the previous system using the BDF or LSODA routines. Comment on the performance of both routines compared with RK23 and RK45.
(c) Plot the results of the numerical integrations of the system of equations, using appropriate axis scales to represent the dynamics of this system.

(7) Consider the following system of ODE representing the decomposition of ozone (Lapidus et al., 1973):

$$\frac{dx}{dt} = -x - xy + \varepsilon\gamma y; \ \frac{dy}{dt} = \frac{x - xy}{\varepsilon} - \gamma y$$

Here, x and y are the dimensionless concentrations of ozone and oxygen, respectively, and the parameters have the values: $\varepsilon = \frac{1}{98}$, $\gamma = 3$.

(a) Integrate the equations over the interval [0, 50] with initial conditions $x(0) = 1$, $y(0) = 0$. Use the RK45 and BDF methods. Graphically compare the results of both methods for each variable separately, using appropriate plots.
(b) Determine the time at which the oxygen concentration reaches a maximum and the value of that maximum, displaying both values. Note: use the option "events" within routine *solve_idp.*

(8) The following differential equation represents the cooling of a solid material when it is exposed to air at temperature: $T_0 = 300$ K (constant):

$$dT/dt = (A/rcV)\{es(T_0^4 - T^4) + h(T_0 - T)\} \quad T(t = 0) = 900$$

where T is the temperature in K, A = 0.25 m^2 is the solid surface, $\rho = 3000$ kg m^{-3} is the solid density, $V = 0.005$ m^3 is the volume, $c = 900$ J kg^{-1} K^{-1} is the specific heat capacity, $h = 30$ J K^{-1} m^{-2} is the natural convection coefficient, $\varepsilon = 0.8$ is the solid emissivity and $\sigma = 5.67 \times 10^{-8}$ W m^{-2} K^{-4} is the Stefan-Boltzmann constant.

(a) Write a script to compute the solution using the RK23 or RK45 routines.
(b) Plot the time profile of the temperature in the solid.
(c) At what time does the temperature reach 305 K? Note: use the option "events" within routine *solve_idp.*

(9) The following differential equations represent fluid motion and heat transfer in a fluid heated from below and subjected to gravity (e.g., the atmosphere), and were described by Lorenz (1963):

$$\frac{dx}{dt} = -P(x - y); \ \frac{dy}{dt} = Rx - y - xz; \ \frac{dz}{dt} = xy - bz$$

where P, R, and b are system parameters: $P = 10$, $R = 28$ and $b = 3$, and the initial condition is: $y_0 = \begin{bmatrix} -7 & 11 & 17 \end{bmatrix}'$

(a) Write a script to compute the solution of the previous equations over the interval $t = 0$ and $t = 10$.
(b) Plot the trajectory followed by the system in the (x,y,z), coordinate system; that is, obtain the "butterfly wing" diagram.
(c) Compare the results after slightly changing the initial condition to $y_0 = [-7.1\ 10.9\ 16.9]'$. Do the results change significantly?

(10) In a closed system, the following irreversible chemical reaction is taking place: $\frac{dC}{dt} = -kC$, and the following experimental data are available: $t_{exp} = [0.2\ 0.5\ 1\ 1.5\ 2]$, $C_{exp} = [0.75\ 0.55\ 0.21\ 0.13\ 0.04]$. Estimate the value of the kinetic rate constant k, and the initial value of the concentration, $C(0)$.

(11) The production of chlorobenzenes is carried out in the liquid phase in a perfectly stirred vessel with a cooling jacket because the reactions are exothermic. In the liquid phase of the vessel, the following chemical reactions occur:

$$C_6H_6 + Cl_2 \xrightarrow{k1} C_6H_5Cl + HCl \qquad k_1 = 510(\text{kgmol/m}_3)^{-1}(\text{h})^{-1}$$

$$C_6H_5Cl + Cl_2 \xrightarrow{k2} C_6H_4Cl_2 + HCl \qquad k_2 = 64(\text{kgmol/m}_3)^{-1}(\text{h})^{-1}$$

$$C_6H_4Cl_2 + Cl_2 \xrightarrow{k3} C_6H_3Cl_3 + HCl \qquad k_3 = 2.1(\text{kgmol/m}_3)^{-1}(\text{h})^{-1}$$

The system includes a reflux that allows hydrochloric acid (HCl) and chlorine vapors to be removed from the system; the amount of chlorine dissolved in the liquid phase is limited by the solubility of chlorine in the liquid solution.

Assume: the system operates under isothermal conditions; changes in the liquid-phase volume are negligible; hydrochloric acid vaporizes and leaves the system; gaseous chlorine dissolves into the liquid solution until it reaches its maximum solubility.

Let N_B, N_M, N_D, and N_T denote the moles of benzene, mono-, di-, and trichlorobenzene, respectively, in the reactor, respectively. The mass balances for each component are as follows (V is the volume of the liquid phase in the reactor):

$$\frac{dN_B}{dt} = -\frac{k_1 N_B N_C}{V}; \quad \frac{dN_M}{dt} = \frac{k_1 N_B N_C}{V} - \frac{k_2 N_M N_C}{V};$$
$$\frac{dN_D}{dt} = \frac{k_2 N_M N_C}{V} - \frac{k_3 N_D N_C}{V}; \quad \frac{dN_T}{dt} = \frac{k_3 N_D N_C}{V}$$

In addition, the mass balance for chlorine dissolved in the liquid phase (N_C) is given by:

$$\frac{dN_C}{dt} = F - \frac{k_1 N_B N_C}{V} - \frac{k_2 N_M N_C}{V} - \frac{k_3 N_D N_C}{V}; \; N_{C,MAX} = 0.12 N_{B,0}$$

where F is the chlorine feed flow rate to the reactor, which is 70 (kmol/h), and $N_{B,0}$ is the initial benzene charge (50 kmol in this case). The liquid-phase volume of the reactor is 73 m^3.

Compute the reaction time required to:

(a) Maximize the production of monochlorobenzene (N_M)
(b) Maximize the production of dichlorobenzene (N_D)

In both cases, your program must return the maximum moles and the corresponding times, and a plot of the simulated moles of the five species as a function of time (in hours).

(12) Here we will consider the following kinetic scheme taking place in a perfectly mixed batch reactor:

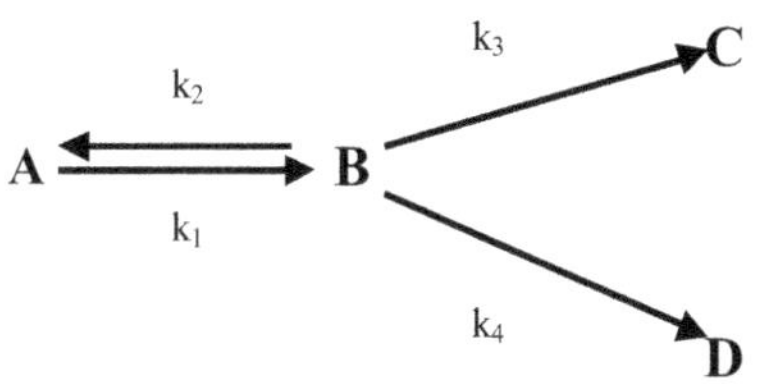

and which is mathematically described by the following system of equations:

$$\frac{dC_A}{dt} = -k_1 \cdot C_A + k_2 \cdot C_B; \quad \frac{dC_B}{dt} = k_1 \cdot C_A - (k_2 + k_3 + k_4) \cdot C_B$$
$$\frac{dC_C}{dt} = k_3 \cdot C_B; \quad \frac{dC_D}{dt} = k_4 \cdot C_B$$

The following data are available:

$$C_{A0} = 50 \text{ gmol/L}; \quad C_{B0} = 5.0 \text{ gmol/L}; \quad C_{C0} = 0 \text{ gmol/L} \quad C_{D0} = 0 \text{ gmol/L}$$
$$k_1 = 2 \text{ h}^{-1}; \quad k_2 = 1 \text{ h}^{-1}; k_3 = 0.2 \text{ h}^{-1}; k_4 = 0.6 \text{ h}^{-1}$$

Calculate the residence time that maximizes the outlet concentration of B.

(13) Consider the following differential equations representing the behavior of a continuous-flow bioreactor with perfect mixing:

$$\frac{dX}{dt} = rk_2C - DX$$
$$\frac{dC}{dt} = k_1S(X - C) - (k_2 + k_3)C - DC$$
$$\frac{dS}{dt} = -k_1S(X - C) + k_3C + D(S_0 - S)$$

where D is the dilution rate; X is the biomass concentration [mol/L]; C is the concentration of a metabolic intermediate (enzyme) [mol/L]; S is the limiting-substrate concentration [mol/L]; and S_0 is the substrate concentration in the feed, which contains neither biomass nor enzyme. The parameters are as follows:

- $r = 0.09$ [mol of biomass formed per mol of enzyme]
- $S_0 = 10$ [mol/L]; $k_1 = 0.9$ [L mol^{-1} h^{-1}]; $k_2 = 0.7$ [h^{-1}]; $k_3 = 0.005$ [h^{-1}]

Find the value of D[h^{-1}] that maximizes the production of the secondary metabolite C; that is, the objective function is $f_{\text{obj}} = D \cdot C$.

Notes:

(a) For the numerical integrations, choose (by trial and error) a sufficiently large time to ensure that the process has reached steady state. For this example, D values range from 0 to 0.55 h^{-1}; the latter corresponds to the bioreactor washout condition.
(b) From the shape of the $C(t)$ curves obtained in step (a), infer how to extract the value of C to be used in the objective function.

(14) Consider the motion of a cork sealing a bottle that contains a pressurized liquid–vapor mixture. The differential equation describing the cork's acceleration is given by (Dordman, 1996):

$$\frac{d^2x}{dt^2} = g(1+q)\left[\left(1+\frac{x}{d}\right)^{-\gamma} + \frac{Rt}{100} - 1 + \frac{qx}{L(1+q)}\right] \quad 0 \le x \le L$$

where $g = 9.81$; $q = 20$; $d = 5$; $\gamma = 1.4$; $R = 4$. L is the length of the bottle neck, which in this case is 3.75.

If the cork is initially at rest, $x(t = 0) = 0$, $dx/dt\ (t = 0) = 0$, determine the time t^* at which the cork leaves the bottle neck, i.e., $x(t^*) = L = 3.75$). What is the cork's velocity at that time? Plot the results for the cork's displacement and velocity as functions of time. Present your results graphically.

References

Bogacki, P., & Shampine, L. F. (1989). A 3(2) pair of Runge-Kutta formulas. *Applied Mathematics Letters, 2*, 1–9.

Butcher, J. C. (1963). On Runge-Kutta processes of higher order. *Journal of Australian Mathmatical Society, 3*, 185–201.

Butcher, J. C. (1964). Implicit Runge-Kutta processes. *Mathematics of Computation, 18*, 50–64. https://www.ams.org/journals/mcom/1964-18-085/S0025-5718-1964-0159424-9/. Accessed 10 January 2026

Butcher, J. (2007). Runge-Kutta methods. *Scholarpedia, 2*(9), 3147. http://www.scholarpedia.org/article/Runge-Kutta_methods. Accessed 10 January 2026.

Caillaud, J. B., & Padmanabhan, L. (1971). An improved semi-implicit Runge-Kutta method for stiff systems. *Chemical Engineering Journal, 2*, 227–232.

Dahlquist, G., & Bjork, A. (1974). *Numerical methods*. Prentice Hall.

Dormand, J. R., & Prince, P. J. (1980). A family of embedded Runge-Kutta formulae. *Journal of Computational and Applied Mathematics, 6*(1), 19–26.

Dormand, J. R. (1996). *Numerical methods for differential equations*. CRC Press.

Fehlberg, E. (1968). Classical fifth-, sixth-, seventh-, and eighth-order Runge-Kutta methods with step size control. NASA Technical Report, NASA TR R-287. https://ntrs.nasa.gov/api/citations/19680027281/downloads/19680027281.pdf. Accessed 10 January 2026

Gear, C. W. (1971). *Numerical initial value problems in ordinary differential equations*. Prentice-Hall.

Gill, S. (1951). A process for the step-by-step integration of differential equations in an automatic computing machine. *Proceedings of the Cambridge Philosophical Society, 47*, 96–108. https://doi.org/10.1017/S0305004100026414. Accessed 10 January, 2026.

Heun, K. (1900). Neue Methoden zur approximativen Integration der Differentialgleichungen einer unabhängigen Veränderlichen. *Zeitschrift fur angewandte Mathematik und Physik, 45*, 23–38.

Hildebrand, F. B. (1987). *Introduction to numerical analysis* (2nd ed.). Dover Books on Mathematics.

Hindmarsh, A. C., & Byrne, G. D. (1976). Applications of EPISODE: An experimental package for the integration of systems of ordinary differential equations. In L. Lapidus & W. E. Schiesser (Eds.), *Numerical methods for differential systems* (pp. 147–166). Academic Press.

Hindmarsh, A. C., & Petzold, L. R. (1995). Algorithms and software for ordinary differential equations and differential-algebraic equations—Part II: Higher-order methods and software packages. *Computers in Physics, 9*(2), 148–155.

Kutta, W. (1901). Beitrag zur näherungsweisen Integration totaler Differentialgleichungen. *Zeitschrift fur angewandte Mathematik und Physik, 46*, 435–453.

Lapidus, L., Aiken, R. C., & Liu, Y. A. (1973). The occurrence and numerical solution of physical and chemical systems having widely varying time constants. In R. A. Willoughby (Ed.), *Stiff differential systems* (pp. 187–200). Plenum.

Lorenz, E. N. (1963). Deterministic nonperiodic flow. *Journal of the Atmospheric Sciences, 20*(2), 130–141.

Luther, H. A., & Konen, H. P. (1955). Some fifth-order classical Runge-Kutta formulas. *SIAM Review, 7*, 551–558.

Luther, H. A. (1966). Further explicit fifth-order Runge-Kutta formulas. *SIAM Review, 8*, 374–380.

Luther, H. A. (1968). An explicit sixth-order Runge-Kutta formula. *Mathematics of Computation, 22*, 434–436.

Michelsen, M. L. (1976). An efficient general purpose method for the integration of stiff ordinary differential equations. *AIChE Journal, 22*, 594–597.

Nakamura, S. (1991). *Applied numerical methods with software*. Prentice Hall.

Petzold, L. R. (1982). Differential/algebraic equations are not ODE. *SIAM Journal on Scientific and Statistical Computing, 3*(3), 367–384.

Petzold, L. R. (1983). Automatic selection of methods for solving stiff and non-stiff systems of ordinary differential equations. *SIAM Journal on Scientific and Statistical Computing, 4*(1), 136–147.

Ralston, A., & Rabinowitz, P. (1978). *A first course in numerical analysis*. McGraw Hill.

Rice, R., & Do, D. D. (2012). *Applied mathematics and modeling for chemical engineering* (2nd ed.). Wiley.

Rosenbrock, H. H. (1963). Some general implicit processes for the numerical solution of differential equations. *Computational Journal, 5*, 329–330. https://doi.org/10.1093/comjnl/5.4.329

Runge, C. (1895). Über die numerische Auflösung von Differentialgleichungen. *Mathematische Annalen, 46*, 167–178.

Shampine, L. F., & Hosea, M. E. (1996). Analysis and implementation of TR-BDF2. *Applied Numerical Mathematics, 20*, 21–37. https://doi.org/10.1016/0168-9274(95)00115-8

Shampine, L. F., & Gordon, M. K. (1975). *Computer solution of ordinary differential equations: The initial value problem.* W. H. Freeman.

Chapter 4
Ordinary Differential Equations: Boundary-Value Problems

4.1 Introduction

In this chapter, we consider the numerical solution of the class of ordinary differential equations (ODE) known as Boundary-Value Problems (BVPs). These numerical problems arise when the system properties cannot be assumed to be uniform and, therefore, spatial variations in concentration, temperature, velocity, etc. appear. Their form depends on how mass, heat, and momentum are exchanged through the boundaries of the system under analysis, and they must be accounted for in the analysis and numerical solution. Thus, these problems originate from transport phenomena. Some examples include:

1. Mass and heat diffusion in a porous solid particle: this type of problem is classical in reactor design and arises because resistance to heat and mass transfer cannot be neglected. In addition, the conditions imposed on the pellet arise from the geometry (heat and mass flux must be zero at its center) and from interaction with the external flow (finite film resistances at the surface). These conditions are specified at different points in the domain where we wish to determine the concentration and/or temperature profile.
2. Steady-state velocity distribution in closed ducts (of any cross-sectional shape). In this case, the pressure gradient in the axial direction is given, and the objective is to determine the velocity profile in the direction transverse to the flow. Here the duct wall represents a sink of momentum, and the fluid velocity must be zero there, which imposes the boundary condition of the problem. It is precisely the boundary conditions that define the nature of the flow (see what happens, for example, when the nozzle of a garden sprinkler is adjusted).

These problems are differential equations, typically second order, with two boundary conditions applied at two distinct points along the spatial coordinate. This feature requires the development of special methods to solve these ODE-BVPs.

H. Jorquera and C. A. Gelmi, *Numerical Methods in Chemical Process Engineering Using Python*, https://doi.org/10.1007/978-3-032-22958-8_4

Here we will consider the most used methods, highlighting their limitations and advantages.

Typically, numerical methods solve for the entire set of discrete values simultaneously, so there is no error propagation in the computation in the sense discussed for ODE initial value problems (IVPs). On the other hand, some amount of error is unavoidable because the methods approximate the desired solution in one way or another. In other words, the phenomenon of error propagation does not play a role here and, therefore, phenomena such as stiffness, for example, do not occur.

4.2 Definition of the Problem

A boundary-value problem of order n with n boundary conditions at equal (or lesser) numbers of points in the x-domain can be written as:

$$y^{(n)} = f\left(x, y, y^{(1)}, \ldots, y^{(n-1)}\right) \tag{4.1}$$

Typically, the conditions are specified at two points in the x-domain, forming what we call a two-point boundary-value problem (PVC).

It is possible to reduce the above n-order equation to a system of n first-order ODE, as was shown for initial value problems (PVI). This enables, in some cases, the use of PVI techniques to solve PVC. Then, instead of (4.1) we can write:

$$y'_i = f_i(x, y_1, y_2, \ldots, y_n); \quad i = 1, 2, \ldots, n \tag{4.2}$$

With n boundary conditions that are not all defined at the same value of x.

Example 4.1 **Conversion of a third-order ODE to a first-order system**

Consider the BVP:

$$\frac{d^3y}{dx^3} + 2y\frac{d^2y}{dx^2} - \left(\frac{dy}{dx}\right)^2 + 1 = 0$$

With boundary conditions:

$$y(0) = 0; \quad \left(\frac{dy}{dx}\right)_{x=0} = 0; \quad \left(\frac{dy}{dx}\right)_{x=2.5} = 1$$

Using the notation: $y_1(x) = y(x)$; $y_2(x) = y'(x)$; $y_3(x) = y''(x)$, we obtain the system:

$$y_1' = y_2; \quad y_2' = y_3; \quad y_3' = (y_1')^2 - 1 - 2 \cdot y_1 \cdot y_3$$
$$y_1(0) = 0; \quad y_2(0) = 0; \quad y_2(2.5) = 1$$

In general, BVP-ODE are solved differently from PVI-ODE, since a complete function must be found as the solution (temperature profile, velocity profile, concentration profile), and not merely a point value, as was done with ordinary differential equations (PVI), in which the solution was advanced step by step estimating the next value of the sought function.

4.3 Most Used Methods

The solution of a BVP-ODE typically consists of reducing the original problem (order n) to a system of linear or nonlinear equations, or to a set of first-order ODE. Subsequently, solving the equations (or the equivalent set of ODE) yields the required function. The most often used methods are:

1. Transformation to a System of Equations: Finite Differences, Finite Elements, Orthogonal Collocations, Weighted Residuals, Variational Methods.
2. Transformation to an Initial Value Problem: Shooting Methods

Here we will concentrate on two methods: finite differences and shooting methods; we will begin by developing the latter, given its similarity to the methods already discussed in the previous chapter.

4.4 Shooting Method

Consider Example 4.1, in which all variables (except y_3) are specified at $x = 0$. Note that if we knew all the values of $y_i(0)$ at $x = 0$, we could obtain all $y_i(x)$ using techniques already discussed for IVPs, i.e., by integrating up to $x = 2.5$ using routines such as RK23, RK45, etc.

Now suppose we estimate a value for $y_3(x = 0)$. Then we could integrate the differential equations from $x = 0$ forward. If the estimated value for $y_3(x = 0)$ is correct, then the boundary condition $y_2(2.5) = 1$ is satisfied, and we have solved the problem by obtaining all the desired profiles $y_i(x)$. This is the basis of the shooting method, by analogy with kinematic problems.

That is, the shooting method can be formulated as an optimization problem for a parameter associated with a system of first-order ordinary differential equations, as discussed in the previous Chapter. The distinction here is that the parameter (or parameters) to be estimated consists of the value(s) that completely define the conditions of all dependent variables at one end of the domain.

In Python, the implementation is presented in Fig. 4.1, and Example 4.2 shows a concrete application.

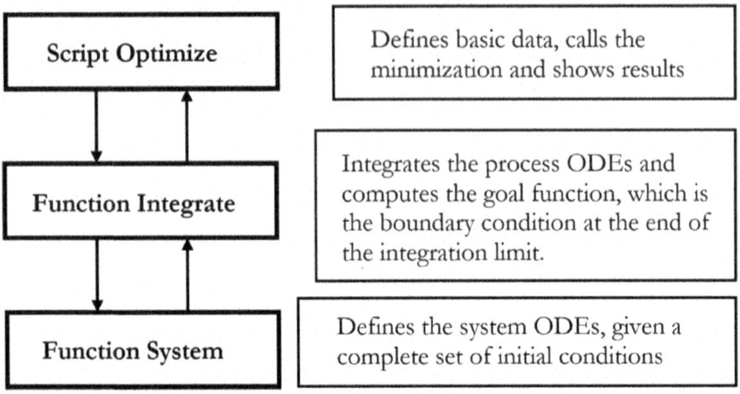

Fig. 4.1 Diagram of solving boundary-value problems in ODE using shooting methods

Example 4.2 Solving ODE by the shooting method

For Example 4.1, the following script can be written:

```
import numpy as np
import matplotlib.pyplot as plt
from scipy.integrate import solve_ivp
from scipy.optimize import minimize

def System(t, y):
    """
    Function that defines the system of ODEs
    """
    yp = np.zeros(3)
    yp[0] = y[1]
    yp[1] = y[2]
    yp[2] = y[1]**2 - 1 - 2*y[0]*y[2]
    return yp

def Integrate(x):
    """
    Computation of the objective function,
    corresponding to the minimization of a boundary condition
    """
    Y0 = np.array([0.0, 0.0, x[0]])

    sol = solve_ivp(
        System,
        (0.0, 2.5),
        Y0,
        method="BDF"   # equivalent to ode23s (stiff solver)
    )

    f = (1.0 - sol.y[1, -1])**2
    return f
```

```
x0 = np.array([1.0])

result = minimize(Integrate, x0)

x_opt = result.x[0]

print("Optimization result for y3(0):")
print(x_opt)

Y0 = np.array([0.0, 0.0, x_opt])

sol = solve_ivp(
    System,
    (0.0, 2.5),
    Y0,
    method="BDF"
)

T = sol.t
Y = sol.y.T

plt.plot(T, Y[:, 0], 'k-', label='y1(x)')
plt.plot(T, Y[:, 1], 'k--', label='y2(x)')
plt.plot(T, Y[:, 2], 'k:', label='y3(x)')

plt.xlabel('x')
plt.ylabel('y1(x), y2(x), y3(x)')
plt.legend(loc='best')
plt.grid(True)
plt.show()
```

Running the script yields the results shown in Fig. 4.2. The value is $y_3(0) = 1.3133$.

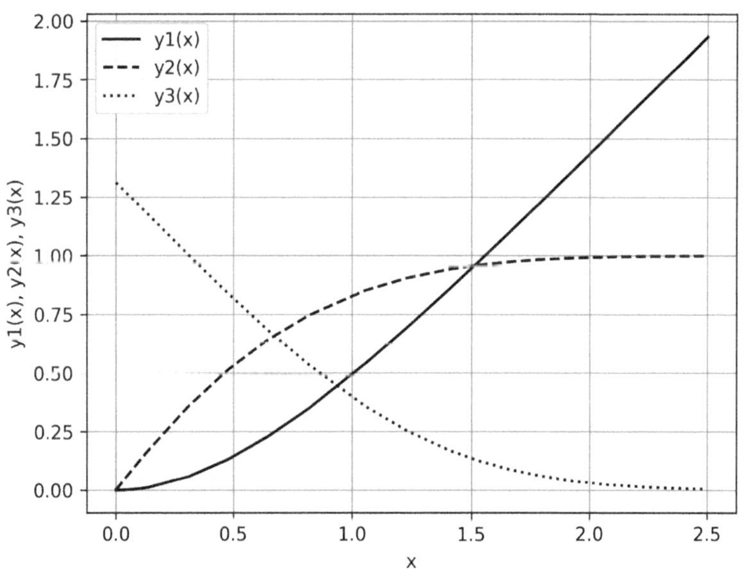

Fig. 4.2 Results of Example 4.2

Example 4.3 **Diffusion and chemical reaction in a catalyst pellet**

Consider the diffusion and chemical reaction problem in a catalyst pellet; since it is a solid, there is no convective term, and the mass balance for a species is (Finlayson, 2012; Levenspiel, 1999; Rice & Do, 2012):

$$\frac{1}{r^{a-1}}\frac{d}{\mathrm{d}r}\left(r^{a-1}\frac{\mathrm{dc}}{\mathrm{d}r}\right) = \Phi^2 f(c); \quad 0 < x < 1;$$

Boundary conditions are:

$$\left(\frac{\mathrm{dc}}{\mathrm{d}r}\right)(r=0) = 0; \quad -\left(\frac{\mathrm{dc}}{\mathrm{d}r}\right)(r=1) = Bi_{\mathrm{M}}(c(r=1)-1); \quad Bi_m = \frac{k_{\mathrm{M}}R}{D_e}$$

The first arises from the symmetry of the profile with respect to the center of the pellet, and the second from the fact that no mass accumulates at the pellet–fluid interface. The parameter Φ^2 corresponds to the Thiele modulus (the ratio of the reaction rate to the diffusion rate), and $\mathrm{Bi_m}$ is the Biot number in mass transfer; the value of a depends on the pellet geometry:

- $a = 1$ for planar geometry
- $a = 2$ for cylindrical geometry
- $a = 3$ for spherical geometry.

First, we reduce the problem to a first-order system; let $y_1 = c(r)$ and $y_2 = \mathrm{d}c/\mathrm{d}r$, which yields the system of equations:

$$\frac{\mathrm{d}y_1}{\mathrm{d}r} = y_2; \quad \frac{\mathrm{d}y_2}{\mathrm{d}r} = \Phi^2 f(y_1) - \frac{(a-1)}{r}y_2; \quad \Phi^2 = \frac{kR^2}{D_e}$$

Boundary conditions are: $y_2(0) = 0$; $y_2(1) = -Bi_m \cdot (y_1(1) - 1)$;

We can integrate the system of ODE, assuming $y_1(0) = p$ (the only parameter that must be defined) from r = 0 to r = 1, which allows the objective function to be defined as follows:

$$f(p) = (y_2(1) + B_{\mathrm{im}} \cdot (y_1(1) - 1))^2$$

If we are going to integrate the system of ODE from $r = 0$ onward, we must avoid division by zero on the right-hand side. Applying L'Hôpital's rule for the spherical geometry ($a = 3$) or cylindrical geometry ($a = 2$) yields:

$$\lim_{r\to 0}\frac{d^2y_1}{dr^2}=\lim_{r\to 0}\left\{\Phi^2 y_1(r)-\frac{a-1}{r}\frac{dy_1}{dr}\right\}$$

$$=\Phi^2 y_1(0)-(a-1)\cdot\lim_{r\to 0}\frac{d^2y_1}{dr^2}$$

$$\rightarrow a\frac{d^2y_1}{dr^2}(0)=\Phi^2 y_1(0)\rightarrow\frac{d^2y_1}{dr^2}(0)=\frac{\Phi^2 y_1(0)}{a}$$

Figure 4.3 shows the results of the shooting method for the spherical geometry case, with $B_{im} = 1$ and the following Thiele modulus values: $\Phi^2 = 0.3, 1, 3, 10$. The initial estimate for the value $y_1(0) = p$ has been taken from the result of the following example.

As Φ^2 increases, there is greater resistance and the internal concentration is lower. When Φ^2 decreases, the diffusive resistance diminishes and the concentration profile becomes flatter and closer to the fluid value (1).

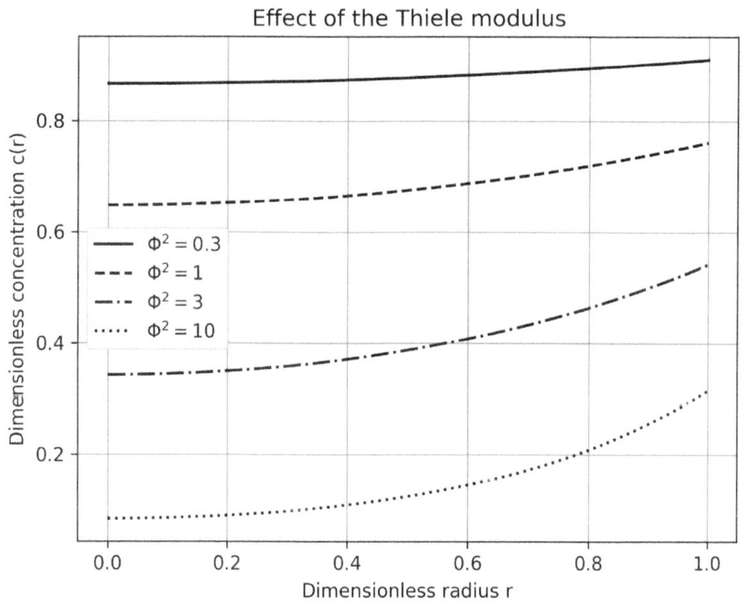

Fig. 4.3 Numerical solution of Example 4.3 using the shooting method ($B_{im} = 1$)

The following script presents the solution of the concentration profile and the generation of Fig. 4.3.

```
import numpy as np
import matplotlib.pyplot as plt
from scipy.integrate import solve_ivp
from scipy.optimize import minimize

def ej4_3(Bim, fis):
    """
    Function that solves Example 4.3
    """
    results = []

    for phi2 in fis:

        # Initial guess for c(0)
        x_ini = Bim / (phi2 / 3.0 + Bim * (1.0 + phi2 / 6.0))

        # Optimization of c(0)
        res = minimize(lambda x: Integrate1(x, phi2), [x_ini])
        x_opt = res.x[0]

        print("Optimization result for c(0):")
        print(x_opt)

        # Verification of the solution
        Y0 = np.array([x_opt, 0.0])

        sol = solve_ivp(
            lambda r, y: System1(r, y, phi2),
            (0.0, 1.0),
            Y0,
            method='RK45',
            t_eval=np.linspace(0,1,50)
        )

        results.append((sol.t, sol.y[0]))

    # Plot results
    linestyles = ['k-', 'k--', 'k-.', 'k:']
    labels = [
        r'$\Phi^2 = 0.3$',
        r'$\Phi^2 = 1$',
        r'$\Phi^2 = 3$',
        r'$\Phi^2 = 10$'
    ]

    for (T, C), style, label in zip(results, linestyles, labels):
        plt.plot(T, C, style, label=label)

    plt.legend(loc='best')
    plt.title('Effect of the Thiele modulus')
    plt.xlabel('Dimensionless radius r')
    plt.ylabel('Dimensionless concentration c(r)')
    plt.grid(True)
    plt.show()

def Integrate1(x, phi2):
    """
    Objective function to be minimized
    """
    Bim = 1.0
    Y0 = np.array([x[0], 0.0])
```

```
    sol = solve_ivp(
        lambda r, y: System1(r, y, phi2),
        (0.0, 1.0),
        Y0,
        method='RK45'
    )

    f = (sol.y[1, -1] + Bim * (sol.y[0, -1] - 1.0))**2
    return f

def System1(r, y, phi2):
    """
    Function that defines the ODE system
    """
    yp = np.zeros(2)
    yp[0] = y[1]

    if r > 0.0:
        yp[1] = phi2 * y[0] - 2.0 * y[1] / r
    else:
        yp[1] = phi2 * y[0] / 3.0

    return yp

Bim=1.0
fis = np.array([0.3, 1.0, 3.0, 10.0])
ej4_3(Bim, fis)
```

4.4.1 Comments on the Shooting Method

(a) Shooting methods allow the use of efficient IVP techniques, e.g., RK23, RK45, BDF, etc.
(b) The method is applicable to both linear and nonlinear problems.
(c) The technique is fast and CPU-efficient.
(d) Although the boundary-value problem is well conditioned (i.e., non-stiff), the system of equations that results from using shooting methods may be stiff, which requires using solvers designed for that purpose, such as BDF, LSODA, etc. Sometimes when the integration in one direction becomes very slow, a possible solution is to integrate in the opposite direction (Sewell, 1988).
(e) Often the numerical solution is highly sensitive to the initial guess for the parameter p, which can sometimes be a vector of parameters. Example 4.4 shows how to choose an initial value for the method.

Example 4.4 **Selection of initial values in shooting methods**

For the previous example, estimate an initial value of $y_1(0) = p$ if $f(c) = c$ and the geometry is spherical.

By performing a Taylor series approximation to the function $y_1(r) = c(r)$, we obtain:

$$
\begin{aligned}
y_1(r) &= y_1(0) + r \cdot \frac{\mathrm{d}y_1}{\mathrm{d}r}(0) + \frac{r^2}{2} \cdot \frac{d^2 y_1}{\mathrm{d}r^2}(0) + \cdots \\
&= p + \frac{r^2}{2} \cdot \frac{d^2 y_1}{\mathrm{d}r^2}(0) + \cdots
\end{aligned}
$$

where we have introduced the definition of p and the boundary condition at $r = 0$ for dy_1/dr. To evaluate the second derivative at $r = 0$ we use the mass balance from previous example, and it is found that the following truncated Taylor approximation for $y_1(r)$ can be used:

$$y_1(r) \approx p \cdot \left\{1 + \frac{r^2\Phi^2}{6}\right\} \Rightarrow y_2(r) = \frac{dy_1(r)}{dr} \approx p \cdot \frac{r \cdot \Phi^2}{3}$$

Replacing this result in the boundary condition to be satisfied at $r = 1$, we obtain:

$$y_2(1) = p \cdot \frac{\Phi^2}{3} = -Bi_m \cdot \left[p \cdot \left\{1 + \frac{\Phi^2}{6}\right\} - 1\right]$$

$$\rightarrow p \approx Bi_m \cdot \left[\frac{\Phi^2}{3} + Bi_m \cdot \left\{1 + \frac{\Phi^2}{6}\right\}\right]^{-1}$$

It is recommended to apply a minimization function (such as *minimize*) using the search interval $p/2$ to 2p to obtain the optimal value $p^* = y_1(0)$.

This methodology can be used for any BVP (linear or nonlinear). The advantage is that it reduces stability and convergence issues of minimization when an initial estimate of p is available.

The reader is left to consider whether this methodology can be extended to the case in which there is a vector of parameters to be optimized via the shooting method.

4.5 Finite Difference Methods

This methodology replaces the derivatives that appear in a boundary-value problem with finite-difference approximations. The result is a system of algebraic equations.

To carry this out, we establish a grid of discrete points over the interval of interest of the independent variable x (Fig. 4.4).

Where a grid has been defined with the following nodes:

$$x_i = a + i \cdot \Delta x;\ 0 \le i \le N;\ \Delta x = (b - a)/N \tag{4.3}$$

Thus, the solution of the algebraic system provides estimates of $y(x)$ at the grid points x_i. Now the (approximate) solution must be computed simultaneously for the entire domain given by the $N + 1$ points $\{x_i\}$.

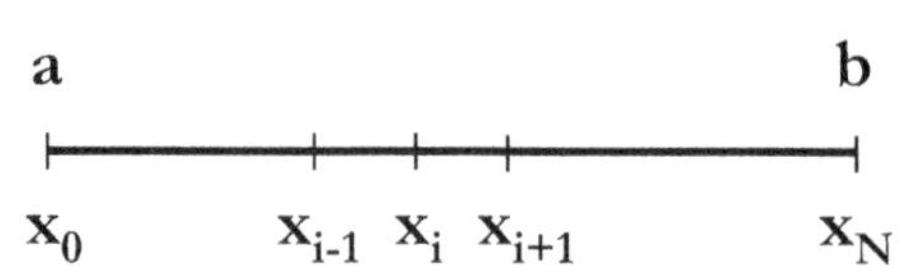

Fig. 4.4 Finite-difference grid for an interval of the independent variable

4.5.1 Finite Differences Approximations

For physical reasons, we will assume that the solution $u(x)$ of our BVP is continuous, so that by writing its Taylor series around $x = x_i$ we have ($h = \Delta x$):

$$
\begin{aligned}
u(x_i + h) &= u(x_i) + hu'(x_i) + \frac{h^2}{2}u(x_i) + \frac{h^3}{6}u'(x_i) + \cdots \\
u(x_i - h) &= u(x_i) - hu'(x_i) + \frac{h^2}{2}u''(x_i) - \frac{h^3}{6}u'''(x_i) + \cdots
\end{aligned} \tag{4.4}
$$

From these expansions, finite-difference approximations to the derivatives of $u(x)$ at various grid positions can be obtained, using the discretized values at those points. Denoting $u_i = u(x_i)$ as the numerical approximations at the grid points, we have the following results:

(a) First-Order Approximations:

$$
\begin{aligned}
u_i' &= \left(\frac{\mathrm{d}u}{\mathrm{d}x}\right)(x = x_i) = \frac{u_{i+1} - u_i}{h} + O(h) \\
u_i' &= \left(\frac{\mathrm{d}u}{\mathrm{d}x}\right)(x = x_i) = \frac{u_i - u_{i-1}}{h} + O(h)
\end{aligned} \tag{4.5}
$$

where the $O(h)$ term indicates that the approximation error decreases linearly as the mesh spacing h decreases.

(b) Second-Order Approximations:

$$
\begin{aligned}
u_i' &= \left(\frac{\mathrm{d}u}{\mathrm{d}x}\right)(x = x_i) = \frac{u_{i+1} - u_{i-1}}{2h} + o(h^2) \\
u_i'' &- \left(\frac{\mathrm{d}^2u}{\mathrm{d}x^2}\right)(x = x_i) = \frac{u_{i+1} - 2u_i + u_{i-1}}{h^2} + o(h^2)
\end{aligned} \tag{4.6}
$$

Similarly, it is possible to construct estimates of higher-order derivatives, as well as higher-accuracy finite-difference approximations.

4.5.2 Construction of the System of Equations

Let us consider in Example 4.5 how the system of equations is constructed to apply finite differences to an ODE boundary-value problem (BVP).

Example 4.5 **Generation of the finite-difference equations**

Consider the following BVP:

$$\frac{d}{\mathrm{d}x}\left(c(u)\frac{\mathrm{d}u}{\mathrm{d}x}\right) = 0; \quad u(0) = 0; \quad u(1) = 1$$

If $c(u) = k = \text{constant.}$, the above equation reduces to $\mathrm{d}^2u/\mathrm{d}x^2 = 0$, and therefore we obtain:

$$\frac{u_{i+1} - 2u_i + u_{i+1}}{h^2} = 0; \quad 1 \le i \le N - 1$$

The system is to be solved together with $u_0 = 0$ and $u_\mathrm{N} = 1$. This gives rise to the following set of $(N + 1)$ linear equations:

$$\begin{bmatrix} 1 & & & & & & & \\ 1 & -2 & 1 & & & & & \\ & 1 & -2 & 1 & & & & \\ & & 1 & -2 & 1 & & & \\ & & & \cdots & & & & \\ & & & & \cdots & & & \\ & & & & & \cdots & & \\ & & & & & & \cdots & \\ & & & & & 1 & -2 & 1 \\ & & & & & & & 1 \end{bmatrix} \begin{bmatrix} u_0 \\ u_1 \\ u_2 \\ u_3 \\ \cdot \\ \cdot \\ \cdot \\ \cdot \\ u_{N-1} \\ u_N \end{bmatrix} = \begin{bmatrix} 0 \\ 0 \\ 0 \\ 0 \\ \cdot \\ \cdot \\ \cdot \\ \cdot \\ 0 \\ 1 \end{bmatrix}$$

Together with the boundary conditions $u_0 = 0$ and $u_\mathrm{N} = 1$, these are reduced to a linear system of $(N - 1)$ equations. This type of tridiagonal structure in the system's matrix is characteristic of all problems solved with second-order central differences in which diffusion or a second spatial derivative appears.

Note that the formulation of the finite-difference problem leads to the requirement to solve the entire system simultaneously, so it is not possible to estimate, for example, u_3, in isolation. The solution of the linear system can be obtained by a direct method such as the Thomas algorithm for tridiagonal matrices (Eq. 1.5) using, for instance, the tridisolve routine presented in Example 1.2 (Moler, 2008). Alternatively, one could use an iterative method for linear systems, such as Gauss–Seidel or successive relaxations (Gerald, 1998; Nakamura, 1991), if N is large, but it should be noted that, in this case, the bands present in the matrix structure must be exploited since the matrix is clearly sparse.

As seen in this Section, the finite-difference formulation of equations for each grid point yields a set of algebraic equations.

4.5.3 General Boundary Conditions

So far, we have discussed boundary-value problems (BVP) in which the boundary conditions are essential, i.e., the value of the dependent variable is known at the interval endpoints. What happens, then, if we have natural boundary conditions that specify the value of the derivative of the dependent variable at (at least) one point in the domain? This type of boundary condition is common in BVP and arises, for example, from walls impermeable to mass or heat diffusion, where the flux must be zero, or from a boundary through which mass or heat is exchanged with the exterior, and the resistance across the interface is known, etc.

To illustrate how to proceed in these more general cases, let us revisit the diffusion and chemical reaction problem in a solid catalyst pellet (Example 4.3).

Expanding the concentration derivatives and using second-order finite differences, we obtain:

$$\frac{c_{i+1} - 2c_i + c_{i-1}}{(\Delta r)^2} + \left(\frac{a-1}{r_i}\right)\frac{c_{i+1} - c_{i-1}}{2\Delta r} = \Phi^2 f(c_i);\ 1 \le i \le N-1 \tag{4.7}$$

Note that in this case we do not know the values C_0 and C_N, which appear in Eq. (4.7) when the index i takes the values 1 and $N-1$, respectively. To obtain these concentrations C_0 and C_N, we use the boundary conditions, together with asymmetric second-order finite-difference expressions, which involve concentration values to the right or to the left of the node, respectively, for C_0 and C_N.

(a) at $r = 0$ we use:

$$\left(\frac{\mathrm{d}c}{\mathrm{d}r}\right)_{(r=0)} = \frac{-3c_0 + 4c_1 - c_2}{2\cdot\Delta r} = 0 \rightarrow -3c_0 + 4c_1 - c_2 = 0 \tag{4.8}$$

(b) Boundary condition at $r = 1$:

$$\left(\frac{\mathrm{d}c}{\mathrm{d}r}\right)_{(r=1)} = \frac{3c_N - 4c_{N-1} + c_{N-2}}{2\cdot\Delta r} = -Bi_M(c_N - 1) \tag{4.9}$$

And rearranging we obtain (in matrix notation):

$$c_{N-2} - 4c_{N-1} + (3 + 2\cdot\Delta r\cdot Bi_M)c_N = 2\cdot\Delta r\cdot Bi_M \tag{4.10}$$

And by combining Eqs. (4.8), (4.9) and (4.10) we obtain a system of $(N+1)$ equations (generally nonlinear) for the values $c_0, c_2, \ldots, c_N$. For the case of first-order or zero-order kinetics, f is linear in c and the respective system becomes linear. Below we show how to solve these equations.

4.5.4 Implementation of the Solution: Fixed-Point Iteration

Equations (4.8), (4.9) and (4.10) must be solved numerically. In general, the system of equations is nonlinear because $f(c)$ is generally nonlinear. In simplified form, the system of equations can be written as:

$$\mathbf{A} \cdot \mathbf{C} = \Phi^2 \cdot (\Delta r)^2 \cdot \mathbf{f}(\mathbf{C}) + \boldsymbol{b} \tag{4.11}$$

where $\boldsymbol{A}$ is the matrix obtained by grouping all terms that are linear in the concentrations at the grid nodes, $\boldsymbol{C} = [C_0, C_1, \ldots, C_N]$ is the vector of unknowns (concentrations at all nodes), $\boldsymbol{f}(\boldsymbol{C})$ is the vector of the function $\boldsymbol{f}(\cdot)$ evaluated at the concentrations C only at interior nodes, and $\boldsymbol{b}$ is the right-hand side resulting from applying the boundary conditions plus whatever remains on the right-hand side of the system of equations. Note that the vector $\boldsymbol{f}(\boldsymbol{C})$ has its first element identically equal to zero, due to the boundary conditions.

Letting $\alpha = \Phi^2 \cdot (\Delta r)^2$, the following sequence of iterative calculations can be proposed as a vector version of the fixed-point method applied to Eq. (4.11):

$$\begin{aligned}
\boldsymbol{C}^{(0)} &= \boldsymbol{A}^{-1} \cdot \boldsymbol{b} \\
\boldsymbol{C}^{(1)} &= \boldsymbol{A}^{-1} \cdot \left\{\alpha * f\left(\boldsymbol{C}^{(0)}\right) + \boldsymbol{b}\right\} \\
\boldsymbol{C}^{(2)} &= \boldsymbol{A}^{-1} \cdot \left\{\alpha * f\left(\boldsymbol{C}^{(1)}\right) + \boldsymbol{b}\right\} \\
&\ldots \\
\boldsymbol{C}^{(k+1)} &= \boldsymbol{A}^{-1} \cdot \left\{\alpha * f\left(\boldsymbol{C}^{(k)}\right) + \boldsymbol{b}\right\}
\end{aligned} \tag{4.12}$$

and this computational sequence may be more robust than obtaining the vector $\boldsymbol{C}$ using some Python optimization routine, such as *minimize*, for example.

Example 4.6 **Application of functional iteration to nonlinear finite differences**

Table 4.1 shows calculations for different geometries and types of function $f(c)$, where the solution has been computed using functional iteration as in Eq. (4.12).

The convergence of the solution as the number of points in the mesh increases depends on the specific problem being solved; therefore, it is recommended to verify

Table. 4.1 Comparison of the numerical solution of Eq. (4.12), implementing fixed-point iteration. In all cases, it is assumed that $Bi_m = 1$, $\Phi^2 = 1$

Spherical geometry and $f(c) = c^2$			Cylindrical geometry and $f(c) = c/(1 + c)$		
N	C_0	C_N	N	C_0	C_N
10	0.6925	0.7941	10	0.6578	0.7737
20	0.7018	0.7987	20	0.6694	0.7803
40	0.7071	0.8013	40	0.6764	0.7841
60	0.7090	0.8022	60	0.6789	0.7854

that the results do not change significantly as the number of grid nodes is further increased. The reader is encouraged to confirm these results by applying the shooting method in each case.

Next, we present the script to compute the solution by functional iteration for spherical geometry and $f(c) = c^2$; the reader is encouraged to modify the code to adapt it to the case of the cylindrical pellet with $f(c) = 1/(1+c)$.

```
import numpy as np

def sphere_it(N, phi2, Bi):
    """
    Function that solves the steady-state conduction problem
    in a solid sphere using functional iteration.
    """

    # Parameter initialization
    dr = 1.0 / N
    R = np.linspace(dr, 1.0, N)
    alpha = phi2 * dr**2

    # Right-hand-side vector (includes boundary conditions)
    b = np.zeros(N + 1)
    b[-1] = 2.0 * Bi * dr

    # Finite-difference matrix (includes boundary conditions)
    A = np.zeros((N + 1, N + 1))

    for i in range(1, N):
        A[i, i] = -2.0
        A[i, i - 1] = 1.0 - dr / R[i]
        A[i, i + 1] = 1.0 + dr / R[i]

    # Boundary condition at r = 0
    A[0, 0] = -3.0
    A[0, 1] = 4.0
    A[0, 2] = -1.0

    # Boundary condition at r = 1
    A[N, N - 2] = 1.0
    A[N, N - 1] = -4.0
    A[N, N] = 3.0 + 2.0 * Bi * dr

    # Chemical reaction term
    f = np.zeros(N + 1)

    # Initial guess for functional iteration
    c0 = np.linalg.solve(A, b)
    k = 0
    error = 1.0

    # Functional iteration loop
    while error > 1e-5 and k < 100:
        f[1:N] = alpha * c0[1:N]**2
        c = np.linalg.solve(A, f + b)
        error = np.linalg.norm(c - c0)
        c0 = c
        k += 1

    return c

N = 10
phi2 = 1.0
Bi = 1.0

c = sphere_it(N, phi2, Bi)
print(' C0 =', c[0])
print('C(N) = ', c[N])
```

4.5.5 *Implementation of the Solution: Newton's Method*

If the fixed-point iteration in Eq. (4.12) does not converge (which can be quickly verified), then Newton's method can be used. Using the notation:

$$\boldsymbol{F}(\boldsymbol{C}) = \boldsymbol{A} \cdot \boldsymbol{C} - \alpha \cdot \boldsymbol{f}(\boldsymbol{C}) - \boldsymbol{b} = \boldsymbol{0} \tag{4.13}$$

the Jacobian of this system of equations is given by:

$$\begin{aligned} \boldsymbol{J} &= \boldsymbol{A} - \alpha \cdot \mathrm{diag}\{0, V_1, V_2, \ldots, V_{N-1}, 0\}; \\ V_i &= \frac{\partial f_i}{\partial C_i}; \quad i = 1, 2, \ldots, N-1 \end{aligned} \tag{4.14}$$

and the call to the minimization function can then be replaced by the Newton iteration sequence:

$$\begin{aligned} \boldsymbol{C}^{(0)} &= \boldsymbol{A}^{-1} \cdot \boldsymbol{b} \\ \boldsymbol{C}^{(1)} &= -\boldsymbol{inv}\left(\boldsymbol{J}^{(0)}\right) \cdot \boldsymbol{F}\left(\boldsymbol{C}^{(0)}\right) \\ \boldsymbol{C}^{(2)} &= -\boldsymbol{inv}\left(\boldsymbol{J}^{(1)}\right) \cdot \boldsymbol{F}\left(\boldsymbol{C}^{(1)}\right) \\ &\ldots \\ \boldsymbol{C}^{(k+1)} &= -\boldsymbol{inv}\left(\boldsymbol{J}^{(k)}\right) \cdot \boldsymbol{F}\left(\boldsymbol{C}^{(k)}\right) \end{aligned} \tag{4.15}$$

This iteration makes it possible to solve Eq. (4.11) with a more robust algorithm (slightly more complex to program) than the algorithm in Eq. (4.12).

Example 4.7 **Application of Newton's method to nonlinear finite differences**

The algorithm in Eq. (4.15) has been applied to the example of the tubular reactor with axial dispersion; see Sect. 4.6. Table 4.2 summarizes the numerical results obtained for the case considered there.

Table 4.2 Numerical results for the finite-difference method applied to the tubular reactor with axial dispersion, with $(kL/uC_0) = 7.5$

$\varepsilon = 0.05^{(\#)}$			$\varepsilon = 0.2$			$\varepsilon = 1$		
N	C_0	C_N	N	C_0	C_N	N	C_0	C_N
10	0.8433	0.1423	10	0.6706	0.1765	10	0.4668	0.2434
20	0.8280	0.1380	20	0.6610	0.1735	20	0.4645	0.2422
40	0.8220	0.1367	40	0.6580	0.1726	40	0.4639	0.2418
60	0.8207	0.1364	60	0.6573	0.1724	60	0.4638	0.2418

$^{(\#)}$ $\varepsilon = DL/u$ (D: axial dispersion coefficient, L, reactor length, u: plug flow rate in the reactor)

The solution converges rapidly as N increases, and for all values of the parameter ε the numerical solution is obtained without difficulty, with only 5 or 6 iterations in each case (for a maximum acceptable error of 10^{-5}).

The script that generates the preceding results is shown below.

```
import numpy as np

def react_disp(alpha, ep, N, tol=1e-5, maxit=100):
    """
    Function that computes the concentration profile in a plug-flow reactor
    with axial dispersion and second-order reaction kinetics.

    Input parameters:
    N     : number of grid points
    tol   : acceptable error tolerance for the solution norm
    maxit : maximum number of Newton iterations allowed
    alpha : k*L/(u*C0), dimensionless kinetic group
    ep    : D*L/u, Peclet number for the flow
            D = axial dispersion coefficient
            L = reactor length
            u = average velocity
            C0 = inlet concentration
            k = second-order kinetic constant

    Output:
    C     : concentration profile
    niter : number of Newton iterations performed
    """

    # Grid spacing
    dx = 1.0 / N

    # Right-hand-side vector (boundary conditions included)
    b = np.zeros(N + 1)
    b[0] = 2.0 * dx / ep

    beta = alpha * dx**2 / ep

    # Finite-difference matrix (including boundary conditions)
    A = np.zeros((N + 1, N + 1))

    for i in range(1, N):
        A[i, i] = -2.0
        A[i, i - 1] = 1.0 + 0.5 * dx / ep
        A[i, i + 1] = 1.0 - 0.5 * dx / ep

    # Boundary condition at inlet (Danckwerts)
    A[0, 0] = 3.0 + 2.0 * dx / ep
    A[0, 1] = -4.0
    A[0, 2] = 1.0

    # Boundary condition at outlet
    A[N, N - 2] = 1.0
    A[N, N - 1] = -4.0
    A[N, N] = 3.0

    # Initialize nonlinear terms
    f = np.zeros(N + 1)
    dJ = np.zeros(N + 1)

    # Initial guess
    C = np.linalg.solve(A, b)

    k = 0
    err = 1.0

    # Newton iteration loop
    while err > tol and k < maxit:
        f[1:N] = beta * C[1:N]**2
        dJ[1:N] = 2.0 * beta * C[1:N]

        F = A @ C - f - b
        err = np.linalg.norm(F)

        J = A - np.diag(dJ)
        C = C - np.linalg.solve(J, F)
```

```
        k += 1

    niter = k
    return C, niter

alpha=7.5
ep=0.05
N=10
C, niter = react_disp(alpha, ep, N)
print("Newton iterations:", niter)
print(' C0 =', C[0])
print('C(N) = ', C[N])
```

4.5.6 *Improving the Accuracy of the Results*

Among the alternatives available, we may cite:

(a) Increase the number of nodes in the mesh: as the mesh is refined, the error incurred when approximating derivatives by finite differences is reduced.
(b) Use higher-order finite differences: this involves deriving approximations of order $O(h^3)$, $O(h^4)$, etc., which leads to sparse matrices with more than three nonzero bands.
(c) Extrapolation (Hanna & Sandall, 1995): let U be the exact solution of the BVP and let $U[h_i]$ be the solution obtained with a mesh spacing h_i. If we use second-order central approximations, the error can be expressed as:

$$U - U[h] = Bh^2 + O(h^4) \tag{4.16}$$

If we solve two different spacings h_1 and h_2, we have:

$$U - U[h_1] = Bh_1^2 + O(h_1^4); \quad U - U[h_2] = Bh_2^2 + O(h_2^4) \tag{4.17}$$

Eliminating B between both equations allows us to obtain the extrapolated solution:

$$U^{ex} = \frac{h_2^2 U(h_1) - h_1^2 U(h_2)}{h_2^2 - h_1^2} + \mathcal{O}(h_1^4) \tag{4.18}$$

For example, in the case where $h_2 = 2h_1$, we obtain:

$$U^{ex} = \frac{4U(h_1) - U(h_2)}{3} + \mathcal{O}(h_1^4) \tag{4.19}$$

Therefore, this extrapolation technique yields numerical approximations of order h^4; that is, the accuracy of the results is significantly improved, at the cost of increasing the number of evaluations of approximate solutions. The reader is encouraged to try this approach with the results of Example 4.7 above and generate a sequence of extrapolated results.

4.5.7 Comments and Conclusions Regarding Finite Differences

(a) These methods are easy to formulate for BVPs in general.
(b) The general procedure consists of formulating the finite-difference equation in the interior of the domain and writing equations for the end nodes using the problem's boundary conditions, employing one-sided finite differences with respect to those end nodes.
(c) To achieve good accuracy, it is necessary to increase the number of points in the mesh, with the consequent increase in computational cost and potential conditioning issues in the resulting matrix.
(d) All derivatives present must be evaluated with the same order of accuracy to prevent significant errors in the solution.
(e) The evaluation of nonlinear terms must be carried out with the same accuracy (or higher) than that used for the derivatives.

4.6 Problems

In principle, the problems presented in this Section (Jorquera & Gelmi, 2014) can be solved using either the shooting method or the finite-difference method. However, in some cases the numerical integration associated with the shooting method produces numerical instability due to stiffness, preventing this methodology from being applied; under such circumstances, the finite-difference method becomes a viable option.

(1) Consider the curing process of a 1 m thick concrete wall, which is subjected to an ambient temperature of 20 °C on both faces (boundary conditions). Assume that the curing process is characterized by a uniform, steady heat generation rate of $\phi = 120\,\mathrm{W\ m^{-3}}$. The concrete has a thermal conductivity $k = 0.3\,\mathrm{W\,m^{-1}\,°C^{-1}}$. Compute the temperature profile within the wall and estimate the maximum temperature reached. Use the shooting method and Python's RK23 integrator. Plot the numerical results. The steady-state heat-conduction equation with constant thermal conductivity k is given by:

$$\frac{d}{\mathrm{d}x}\left(k\frac{\mathrm{d}T}{\mathrm{d}x}\right) = k\left(\frac{\mathrm{d}^2T}{\mathrm{d}x^2}\right) = -\phi; \quad T(x=0) = 20; \quad T(x=1) = 20$$

Introducing the definitions: $y_1(x) = T(x); y_2(x) = \frac{dT(x)}{dx}$, the original problem can be written as follows:

$$\frac{dy_1}{dx} = y_2; \quad \frac{dy_2}{dx} = -\frac{\phi}{\boldsymbol{k}}; \quad y_1(x=0) = 20; \quad y_1(x=1) = 20;$$

Solve the problem using the shooting method, iterating on the value of $y_2(x=0)$ and integrating the differential equations until finding the value of $y_2(x=0)$ that makes $y_1(x=1) = 20$.

(2) Consider a tubular reactor in which the degradation of a certain contaminant occurs. The flow in the reactor is of the plug-flow type, that is, with a uniform velocity profile. In addition, axial dispersion exists in the reactor. The degradation rate is second order. The concentration profile within the reactor is given by the solution of the equation:

$$\varepsilon\frac{d^2\phi}{dx^2} - \frac{d\phi}{dx} - \left(\frac{kL}{uC_0}\right)\phi^2 = 0; \quad 0 < x < 1$$

where

- $\varepsilon = D/(Lu)$; D = axial dispersion coefficient.
- L = reactor length.
- u = fluid velocity in the reactor.
- $x = z/L$ = distancia adimensional a lo largo del reactor.
- $\phi = C(x)/C_0$ = dimensionless concentration, where C_0 is the upstream concentration of the reactor.
- k = the kinetic rate constant for degradation of the substance.

The preceding equation is subject to the boundary conditions:

$$\varepsilon\left(\frac{d\phi}{dz}\right)(x=0) = \phi(0) - 1; \quad \left(\frac{d\phi}{dz}\right)(x=1) = 0$$

Defining $y_1(x) = \phi(x)$ e $y_2(x) = \phi'(x)$, one obtains:

$$\frac{dy_1}{dx} = y_2; \quad \frac{dy_2}{dx} = \frac{1}{\varepsilon}\left\{y_2 + \left(\frac{kL}{uC_0}\right)(y_1)^2\right\}$$

The boundary conditions can now be written as:

$$y_2(x=0) = \frac{y_1(0) - 1}{\varepsilon}; \quad y_2(x=1) = 0$$

You are asked to compute the concentration profiles $\phi(x)$ for the cases $\varepsilon = 0.05, 0.2$, and 1.0, that is, progressively increasing the axial-dispersion rate along the reactor. In all cases, assume that the parameter $(kL/uC_0) = 7.5$.

To compute the profiles, call the ODE integrator again using the optimal value of $y_1(0)$ returned by the optimization routine in each case.

(3) When analyzing incompressible, steady flow over a flat plate, one arrives at the following ordinary differential equation to obtain the velocity profile:

$$f\frac{d^2f}{dz^2} + 2\frac{d^3f}{dz^3} = 0$$

Subject to the following boundary conditions:

$$f(0) = 0; \quad \frac{df}{dz}(z = 0) = 0; \quad \frac{df}{dz}(z = Z) = 1$$

(a) Use the shooting method to determine the value of d^2f/dz^2 at $z = 0$. Use different values of the right boundary limit Z, increasing it from 4 to 10, until the value of d^2f/dz^2 at $z = 0$ no longer changes significantly. Then plot the solution for the dimensionless velocities U and V.

$$U(z) \equiv \frac{u(x, y)}{u_\infty} = f'(z); \; z \equiv y\sqrt{\frac{u_\infty}{\nu x}};$$
$$V(z) \equiv \frac{v(x, y)}{\sqrt{\nu u_\infty / x}} = 2\big(z \cdot f'(z) - f(z)\big)$$

(b) Can you solve this problem by applying the finite-difference method? Explain how you would carry out the numerical implementation in this case.

(4) One approach used to improve heat removal from electronic equipment is to add external cooling fins to one of the system surfaces. For a cylindrical fin, the steady-state energy balance yields the ordinary differential equation:

$$k\frac{d}{dx}\left(A\frac{dT}{dx}\right) = hP(T - T_A)$$

where A and P are the fin cross-sectional area and perimeter ($\pi D^2/4$ and πD, respectively, with D the fin diameter). The other terms appearing in the preceding equation are k(the fin thermal conductivity), h (the air-side convection coefficient), and T_A is the ambient temperature. The boundary conditions are given by:

$$T(x = 0) = T_0; \quad \frac{dT}{dx}(x = L) = 0$$

where L is the total fin length. Solve the preceding equation using the shooting method and obtain the temperature profiles for the following conditions:

$k = 1.5\,\text{BTU}\,\text{h}^{-1}\,{}^{\circ}\text{F}^{-1}\,\text{pulg}^{-1}$ $L = 2''$ $h = 0.2\,\text{BTU}\,\text{h}^{-1}\,{}^{\circ}\text{F}^{-1}\,\text{pulg}^{-2}$
$T_0 = 200\,{}^{\circ}\text{F}$ $T_A = 80\,{}^{\circ}\text{F}$

(5) Consider the problem of heat and mass diffusion within a flat pellet of a porous solid (catalyst); the corresponding equations (for first-order kinetics) are:

$$\frac{\mathrm{d}^2y}{\mathrm{d}x^2} - \phi^2 \cdot y \cdot \exp\left(\frac{\gamma\theta}{1+\theta}\right) = 0; \quad \left(\frac{\mathrm{d}y}{\mathrm{d}x}\right)(x=0) = 0; \quad y(x=1) = 1$$

$$\frac{\mathrm{d}^2\theta}{\mathrm{d}x^2} + \beta \cdot \phi^2 \cdot y \cdot \exp\left(\frac{\gamma\theta}{1+\theta}\right) = 0; \quad \left(\frac{\mathrm{d}\theta}{\mathrm{d}x}\right)(x=0) = 0; \quad \theta(x=1) = 1$$

where y and θ correspond to the dimensionless concentration and temperature, respectively. The parameters ϕ^2, γ, and β correspond, respectively, to a dimensionless reaction-rate parameter, a dimensionless activation energy, and a dimensionless heat of reaction. A quantity of interest is the effectiveness factor η, which indicates how effectively the catalyst is utilized; this factor is defined by the expression:

$$\eta = \frac{1}{\phi^2}\left(\frac{\mathrm{d}y}{\mathrm{d}x}\right)(x=1)$$

Write the scripts needed to solve this problem using the finite-difference method. Consider the following values: $\phi^2 = 10$, $\beta = 25$, $\gamma = 5$; your program must also plot the results for both profiles $(y(x), \theta(x))$ and the value of η.0.3

(6) Consider the Graetz problem for laminar heat transport between two flat plates. Here, the eigenvalue problem must be solved:

$$\frac{\mathrm{d}^2y}{\mathrm{d}x^2} + \lambda\left(1 - x^2\right)y = 0$$

Subject to:

$$y'(0) = 0; \quad y(1) = 0$$

That is, one must compute the (positive) eigenvalue λ that satisfies the preceding equation along with the two boundary conditions mentioned above.

(a) First, rewrite the preceding problem as a system of first-order differential equations by defining $y_1(x) = y(x)$ and $y_2(x) = y'(x)$, but now define $y_3(x) = \lambda$, so that the system becomes:

$$\frac{\mathrm{d}y_1}{\mathrm{d}x} = y_2; \quad \frac{\mathrm{d}y_2}{\mathrm{d}x} = -y_1y_3\left(1 - x^2\right); \quad \frac{\mathrm{d}y_3}{\mathrm{d}x} = 0$$

Subject to the boundary conditions $y_1(1) = 0$ and $y_2(0) = 0$, and the normalization condition $y_1(0) = 1$.

(b) Now the boundary condition for $y_1(x)$ at $x = 0$ and the value of the parameter λ must be incorporated as unknowns to be solved for, since neither value is known as a priori. These boundary conditions will determine the numerical solution of the system of equations for $y_1(x)$ and $y_2(x)$ over the entire domain (y_3 remains trivially equal to λ at all times). In particular, the numerical solution for $y_1(x)$ at $x = 1$ will be obtained, which by the boundary condition must be equal to 0. This can be cast as a shooting method with two unknown boundary conditions, that is, an objective function that depends on two values.

(c) Finally, call *minimize* to determine the value of the condition $y_1(0)$ and of λ such that $y_1(1) = 0$ is satisfied.

You are asked to compute the first three eigenvalues and the corresponding solutions for the profile $y(x)$, which should be plotted together with $y'(x)$ to verify that both boundary conditions are satisfied.

(7) Consider the case of two gases, A and B, diffusing through a stagnant film of a gas C (e.g., air) of length $L = 0.001$ m; there are no chemical reactions among the gases, and mass transport occurs only by the mechanism of molecular diffusion. Because this is a three-gas mixture, the appropriate equations to be solved are the following:

$$\frac{dC_A}{dz} = \frac{(x_A \cdot N_B - x_B \cdot N_A)}{D_{AB}} + \frac{(x_A \cdot N_C - x_C \cdot N_A)}{D_{AC}}$$
$$\frac{dC_B}{dz} = \frac{(x_B \cdot N_A - x_A \cdot N_B)}{D_{AB}} + \frac{(x_B \cdot N_C - x_C \cdot N_B)}{D_{BC}}$$
$$\frac{dC_C}{dz} = \frac{(x_C \cdot N_A - x_A \cdot N_C)}{D_{AC}} + \frac{(x_C \cdot N_B - x_B \cdot N_C)}{D_{BC}}$$

where x_i is the mole fraction of gas i, defined by $x_i = C_i/C_T$, where the total concentration is $C_T = C_A + C_B + C_C =$ constant (since there are no reactions). The coefficients D_{ij} represent the diffusivity of gas i in gas j. In addition, the following data are known for the concentrations at the ends of the film and for the diffusion coefficients:

Gas component	Concentration at $z = 0$ kgmol/m^3	Concentration at $z = L$ kgmol/m^3	Diffusion coefficient, m^2/s
A	2.23×10^{-4}	0	$D_{AC} = 1.08 \times 10^{-4}$
B	0	2.70×10^{-3}	$D_{BC} = 1.25 \times 10^{-4}$
C	7.21×10^{-3}	4.73×10^{-3}	$D_{AB} = 1.47 \times 10^{-4}$

The molar fluxes N_i of each gas i are constant quantities (measured in kgmol/(m^2 s)), but their values are unknown, except that $N_C = 0$ because gas C is stagnant.

Use the information in the Table above and the profile equations to:

(a) Compute the parameters N_A and N_B.
(b) Plot the mole fractions x_A, x_B, and x_C as functions of z, using appropriate titles and legends.

Note: for the initial estimate of the parameters N_A and N_B, assume the binary-diffusion simplification:

$$N_A^{(0)} = -D_{AC} \cdot \frac{C_A(z = L) - C_A(z = 0)}{L};$$

$$N_B^{(0)} = -D_{BC} \cdot \frac{C_B(z = L) - C_B(z = 0)}{L}$$

(8) In the process of forming a synthetic fiber, a molten polymer is pressure-driven through an orifice plate and subjected to tension at the thinner end, where a roller continuously draws off the stretched fiber (Middleman, 1977). The equation describing how the material-flow velocity U varies along the forming direction x is given by:

$$\frac{d^2U}{dx^2} = \frac{\rho}{3 \cdot \mu} \cdot U \cdot \frac{dU}{dx} + \frac{1}{U} \cdot \left(\frac{dU}{dx}\right)^2$$

where ρ and μ are the density and viscosity of the polymeric material, respectively. The preceding equation is subject to the boundary conditions:

$$U(x = 0) = U_0;\ U(x = L) = U_L$$

Construct a function that takes as inputs the length L, the polymer density and viscosity (ρ and μ), and the fiber velocities at $x = 0$ and $x = L$, and that plots the solution for the velocity $U(x)$, solving it using the shooting method.

The function must also produce a plot of the fiber radius R as a function of distance x, where R is defined by the continuity equation:

$$\pi(R(x))^2 \cdot U(x) = \pi(R(x = 0))^2 \cdot U(x = 0) = \pi(R(x = L))^2 \cdot U(x = L)$$

Test your routine with the following data:
$L = 300\,\text{cm}$, $U_0 = 15\,\text{cm/s}$, $U_L = 1500\,\text{cm/s}$, $\rho = 0.96\,\text{g/cm}^3$ and $\mu = 5000\,\text{poise}$.

(9) The figure shows a stagnant liquid film (of thickness L) in contact with a gas phase, in which the following irreversible reaction takes place: $A + B \rightarrow C$, where A is a gas that diffuses into the liquid film, dissolves, and reacts with compound B in the liquid phase to produce compound C, with a reaction rate given by $r_C = k\,C_A\,C_B$. The only transport mechanism for A, B, and C in the stagnant liquid film is diffusion.

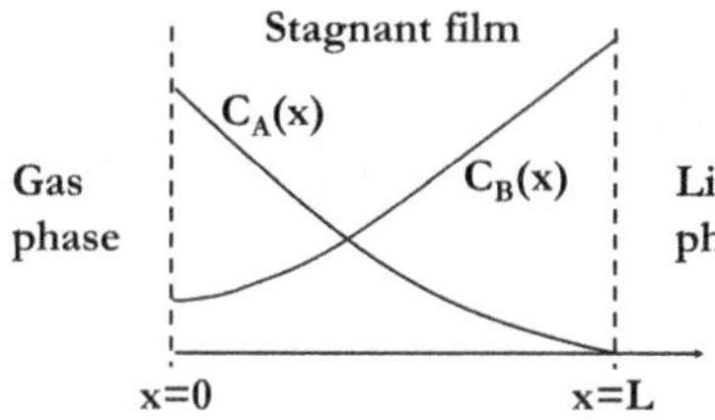

On the other side of the stagnant film is the liquid phase, which is assumed to be well mixed and where the concentration of A is essentially zero. Compound B cannot be transferred to the gas phase because it is only slightly volatile. The equations defining the concentration profiles $C_A(z)$ and $C_B(z)$ are given by:

$$\frac{\mathrm{d}^2 C_A}{\mathrm{d}x^2} = \frac{k \cdot C_A \cdot C_B}{D_{AL}}; \quad \frac{\mathrm{d}^2 C_B}{\mathrm{d}x^2} = \frac{k \cdot C_A \cdot C_B}{D_{BL}}$$

The preceding equations are subject to the following boundary conditions:

$$x = 0: \quad C_A = C_{A0}, \quad \frac{\mathrm{d}C_B}{\mathrm{d}x} = 0; \quad x = L: \quad C_A = 0, \quad C_B = C_{BL}$$

Write a function that uses the finite-difference method to compute the concentration profiles C_A and C_B as functions of the distance x within the stagnant liquid layer, plot both profiles (in the same figure but in separate subplots, with titles and legends), and plot in a second, separate figure the molar fluxes of both components on the same graph (in the same figure, with titles and legends), which are given by:

$$N_A = -D_{AL} \cdot \frac{\mathrm{d}C_A}{\mathrm{d}x}; \quad N_B = -D_{BL} \cdot \frac{\mathrm{d}C_B}{\mathrm{d}x}$$

Test your function with the following data:

$k = 1.6 \times 10^{-3}\,\mathrm{m^3/(s\,kgmol)}$, second – order kinetic constant
$D_{AL} = 2 \times 10^{-10}\,\mathrm{m^2/s}$, diffusivity of A in the liquid
$D_{BL} = 4 \times 10^{-10}\,\mathrm{m^2/s}$, diffusivity of B in the liquid
$L = 2 \times 10^{-4}\,\mathrm{m\,cm}$, stagnant film thickness
$C_{AS} = 1\,\mathrm{kgmol/m^3}$, concentration of A on the surface of the liquid
$C_{BL} = 1.5\,\mathrm{kgmol/m^3}$, concentration of B within the liquid

(10) Consider heat conduction through the wall of a furnace, where the process can be assumed to occur in one dimension. Because there is no internal heat generation within the wall, the heat flux (q_x/A) must be constant; therefore, Fourier's law in the x-direction can be written as:

$$\frac{\mathrm{d}T}{\mathrm{d}x} = -\frac{(q_x/A)}{k}$$

In addition, because there are large temperature variations across the wall, the thermal conductivity k(W/m/K) cannot be considered constant; instead, it varies with temperature according to the equation (with T in K): $k = 30\,(1 + 0.002\,T)$.

The boundary conditions for this problem are:

(a) In $x = 0$, $T(x = 0) = T_0 = 290\,\mathrm{K}$
(b) In $x = L$,

$$\left(\frac{q_x}{A}\right)(x = L) = \sigma \cdot \left\{T_S^4 - T_H^4\right\};\ T_S = T(x = L);$$
$$T_H = 1273K;\ \sigma = 5.68 \times 10^{-8}\frac{W}{m^2K^4}$$

If the wall thickness is $L = 0.25$ m, compute and plot the temperature profile $T(x)$ across the furnace wall, determining the inner-surface temperature T_S and the heat flux (q_x/A) leaving the furnace.

(11) In a cylindrical reactor with plug flow (velocity, temperature, and concentration uniform in the radial direction) and axial dispersion, the concentration and temperature profiles along the reactor satisfy the following ordinary differential equations in dimensionless variables (Finlayson, 2012):

$$\frac{1}{\mathrm{Pe}}\frac{\mathrm{d}^2c}{\mathrm{d}x^2} - \frac{\mathrm{d}c}{\mathrm{d}x} = R(c, T);\quad \text{subject to :}$$
$$-\frac{1}{\mathrm{Pe}}\frac{\mathrm{d}c}{\mathrm{d}x}(0) = 1 - c(0);\ \frac{\mathrm{d}c}{\mathrm{d}x}(1) = 0$$
$$\frac{1}{\mathrm{Pe_H}}\frac{\mathrm{d}^2T}{\mathrm{d}x^2} - \frac{\mathrm{d}T}{\mathrm{d}x} = -\beta \cdot R(c, T)\quad \text{subject to :}$$
$$-\frac{1}{\mathrm{Pe_H}}\frac{\mathrm{d}T}{\mathrm{d}x}(0) = 1 - T(0);\ \frac{\mathrm{d}T}{\mathrm{d}x}(1) = 0$$
$$R(c, T) = 4 \cdot c^2 \cdot \exp\{\gamma - \gamma/T\}$$

Write a function that takes the parameters $\{\mathrm{Pe}, \mathrm{Pe_H}, \beta, \gamma\}$ as inputs and returns the temperature and concentration profiles along the reactor ($0 \le x \le 1$). The routine must also plot both profiles in the same figure (but in separate subplots), including appropriate axis titles, and must also display the corresponding input parameters in the plot. Use the finite-difference method. Test your routine with the following parameter combinations.

(i) $\mathrm{Pe} = 100$; $\mathrm{Pe_H} = 100$; $\beta = 0.1$; $\gamma = 10$
(ii) $\mathrm{Pe} = 5$; $\mathrm{Pe_H} = 15$; $\beta = 0.3$; $\gamma = 10$

(12) The following equation corresponds to a model of a spontaneous-combustion process (Davis, 1962):

$$\frac{d^2y}{dx^2} + \lambda \cdot \exp(y) = 0;\; y(x=0) = 0;\; y(x=1) = 0$$

For this equation, it has been shown that two distinct solutions for $y(x)$ exist when the condition $0 \leq \lambda < \lambda^* = 3.51383$ is satisfied.

Construct a function that solves the above problem using the finite-difference method, finds both solutions, and plots them (in separate plots), including appropriate titles and legends. Test your function for the case $\lambda = 1$.

Note: for $\lambda = 1$, try one solution with initial values that are positive but less than 0.2, and a second solution with initial values that are positive but larger than in the first solution.

(13) The following differential equation describes the velocity of a viscous liquid film discharged from a thin slot:

$$\frac{d^2y}{dx^2} - \frac{1}{y} \cdot \left(\frac{dy}{dx}\right)^2 - y \cdot \frac{dy}{dx} = -1$$

where y is the dimensionless velocity and x is the distance from the slot. This equation is subject to the following boundary conditions:

$$y(x=0) = 0.325 \quad \lim_{x\to\infty}\left(\frac{dy}{dx}\right) = \frac{1}{\sqrt{2 \cdot x}}$$

Use the finite-difference method to solve this problem. To do so, first determine what right-hand integration limit is sufficiently large to satisfy the second boundary condition.

(14) The following differential model (Seydel, 1989) describes a mechanism for the propagation of nerve impulses:

$$\frac{dy_1}{dx} = 3 \cdot T \cdot \left(y_1 + y_2 - \frac{(y_1)^3}{3} - 1.3\right); \quad 0 < x < 1$$

$$\frac{dy_2}{dx} = -\frac{T}{3} \cdot (y_1 - 0.7 + 0.8 \cdot y_2); \quad 0 < x < 1$$

Here, T is the period of the mechanism, and the variable $x = t/T$ makes it possible to formulate the problem with the following boundary conditions (Shampine et al., 2003):

$$y_1(x=0) = 1; \quad y_1(x=1) = 1; \quad y_2(x=0) = y_2(x=1);$$

Solve this problem using the shooting method. Determine the value of T and plot the results on the time scale $t = x \cdot T$ from 0 to $3 \cdot T$, to verify that the numerical solutions for y_1 and y_2 are indeed periodic. It is recommended to start with an initial estimate $T = 2\pi$.

(15) Consider a porous membrane coated with a catalyst, supported on a membrane that is impermeable to diffusion. On the liquid side, a solute diffuses into the membrane pores, where it is degraded by a first-order photocatalytic reaction. The membrane is irradiated from the liquid side, and within it the radiant flux is attenuated by the liquid's turbidity as it penetrates the pores. Under steady-state conditions, the equation describing the solute concentration in the membrane pores is given, in dimensionless form, by Edwards et al. (1996):

$$\frac{\mathrm{d}^2 c}{\mathrm{d}x^2} = \phi^2 \cdot c \cdot \exp(-OD \cdot x);\ 0 < x < 1$$

where

- $c = \frac{C_S}{C_{S0}}$ is the dimensionless solute concentration, referenced to its value at $x = 0$.
- $x = \frac{z}{L}$ is the dimensionless distance, where L is the pore length (or membrane thickness).
- ϕ^2 is the Thiele modulus, $\phi^2 = 2L^2 r_0/(D\, C_{S0}\, r_m)$, where r_0 is the reaction rate at $x = 0$, D is the solute diffusivity in the liquid, and r_m is the average pore radius.
- OD is the optical thickness in the liquid (constant).

The boundary conditions for the preceding equation are:

$$c(x = 0) = 1;\ \frac{\partial c}{\partial x}(x = 1) = 0;$$

Write a function that uses the finite-difference method to solve the preceding equations, plotting $c(x)$ on the same figure for each of the following cases: $OD = 0.1$, 1, and 5, with $\phi^2 = 1$.

References

Davis, H. T. (1962). *Introduction to nonlinear differential and integral equations*. Dover.

Edwards, M. E., et al. (1996). Effectiveness factors for photocatalytic reactions occurring in planar membranes. *Industrial Engineering Chemistry Research, 35*(3), 712–720.

Finlayson, B. (2012). *Introduction to chemical engineering computing* (2nd ed.). Wiley.

Gerald, C. F. (1998). *Applied numerical analysis* (6th ed.). Addison-Wesley.

Hanna, O. T., & Sandall, O. C. (1995). *Computational methods in chemical engineering*. Prentice Hall International Series.

Jorquera, H., & Gelmi, C. (2014). Métodos Numéricos Aplicados a Ingeniería. Casos de estudio usando MATLAB®. Ediciones UC, Santiago, Chile.

Levenspiel, O. (1999). *Chemical reaction engineering* (3rd ed.). Wiley.

Middleman, S. (1977). *Fundamentals of polymer processing*. McGraw-Hill.

Moler, C. (2008). *Numerical computing with MATLAB.* http://www.mathworks.com/moler/. Accessed: 10 January, 2026.

Nakamura, S. (1991). *Applied numerical methods with software.* Prentice Hall.

Rice, R. G., & Do, D. E. (2012). *Applied mathematics and modeling for chemical engineering* (2nd ed.). Wiley.

Sewell, G. (1988). *The numerical solution of ordinary and partial differential equations.* Academic Press.

Shampine, L. F., Gladwell, L., & Thompson, S. (2003). *Solving ODEs with MATLAB.* Cambridge University Press.

Seydel, R. (1989). *From equilibrium to chaos. Practical bifurcation and stability analysis.* Elsevier.

Chapter 5
Partial Differential Equations (PDE)

5.1 Introduction

The distinctive feature of PDE is that the sought solution depends on at least two independent variables, one of which may be time. Classical examples include transient mass and/or heat transfer in solids, or the steady-state distribution of concentrations, velocities, or temperatures in more than one spatial dimension, liquid flow over the surface of a solid, and so forth.

Many numerical techniques have been developed to solve PDE. Initially, finite-difference techniques were the most widely used due to their ease of implementation. However, over the last 30 years, finite element methods have been developed to address problems with complex geometries of more realistic models. In chemical reactor analysis, orthogonal polynomial approximation techniques have been used for reasons of efficiency and accuracy. Another popular approach is the method of lines, which transforms a PDE into a system of ODE that can be solved with efficient algorithms such as those presented in Chap. 3.

Selecting a particular method is not straightforward. It often becomes a matter of experimentation, since many factors must be weighed. These include the mathematical complexity of the problem, the simplicity of implementing the proposed method, the required accuracy, stability issues, and computational cost.

Next, we will present several classical examples of PDE, as well as the most used numerical methods for their solution.

H. Jorquera and C. A. Gelmi, *Numerical Methods in Chemical Process Engineering Using Python*, https://doi.org/10.1007/978-3-032-22958-8_5

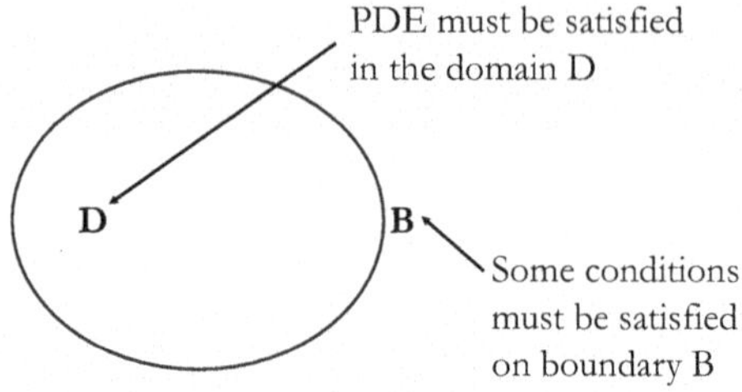

Fig. 5.1 Domain and boundary conditions for equilibrium problems

5.1.1 Equilibrium Problems

Here, the objective is to obtain the solution of a PDE in a closed domain subject to a given set of boundary conditions. Therefore, equilibrium problems are boundary-value problems in more than one dimension. Examples include steady-state distributions of temperatures and concentrations, ideal incompressible flows, stress distributions in solids, and so forth. In this case, the solution of the PDE at each point in the domain depends on the boundary conditions prescribed on the boundary B (Fig. 5.1).

Example 5.1 The Laplace equation for spatial temperature distributions

The steady-state temperature distribution in a solid medium is governed by the Laplace equation. A typical problem is to determine the temperature distribution in a two-dimensional solid whose boundaries are maintained at constant temperature. Therefore, one must solve:

$$\nabla^2 T = \frac{\partial^2 T}{\partial x^2} + \frac{\partial^2 T}{\partial y^2} = 0; \quad 0 < x < 1; 0 < y < 1.$$

With boundary conditions:

$$T(0, y) = 0; \quad T(x, 0) = T_\mathrm{w}; \quad T(1, y) = 0; \quad T(x, 1) = 0.$$

For this equation, an analytical solution exists (Weinberger, 1995):

$$T(x, y) = 2T_w \sum_{n=1}^{\infty} \left[(-1)^n - 1\right] \cdot \frac{\sin(n \cdot \pi \cdot x)}{n \cdot \pi} \cdot \frac{\sinh\left[n \cdot \pi \cdot (y - 1)\right]}{\sinh(n \cdot \pi)}.$$

The solution at each point (x, y) in the domain D depends on the specific conditions on the boundary B (Fig. 5.2). This feature is fundamental and common to all equilibrium problems. However, since the convergence of the numerical series in the preceding equation is unknown *a priori*, it is not obvious how efficient the above formula is for computing the solution $T(x, y)$ at an arbitrary point (x, y) in the domain D.

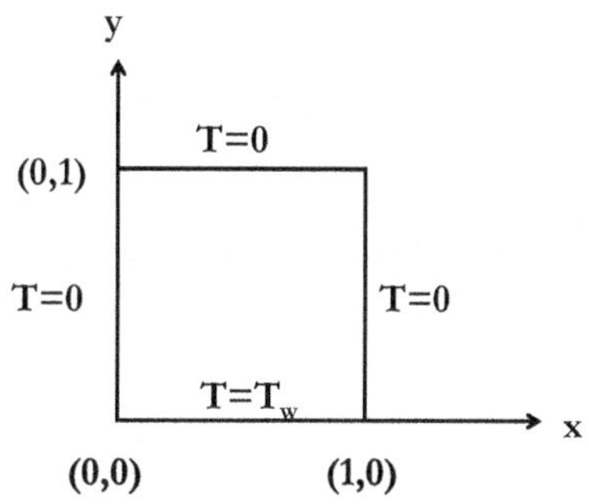

Fig. 5.2 Domain and boundary conditions for the Laplace equation

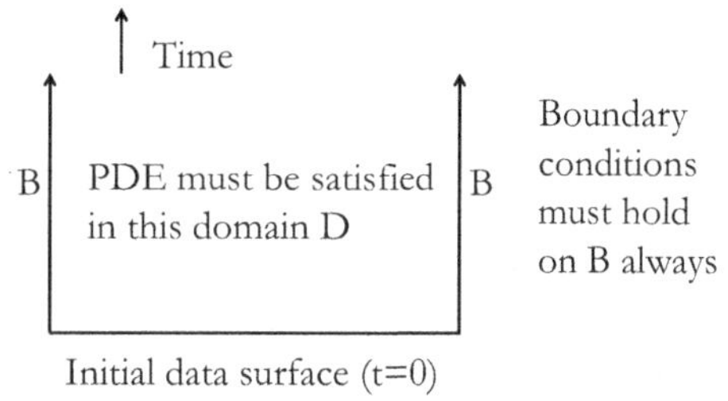

Fig. 5.3 Example of a propagation problem

5.1.2 Propagation Problems

These are transient phenomena in which the solution of the PDE is required in an open domain subject to a set of initial and boundary conditions. This category corresponds to initial-value problems or to initial- and boundary-value problems. The solution must be computed by marching forward in time from the initial data surface, while always satisfying the boundary conditions (Fig. 5.3).

Example 5.2 The Fourier equation for transient temperature distributions

The transient temperature distribution in a one-dimensional solid with thermal diffusivity α, if its initial temperature was uniform and equal to 0 °C and, for $t > 0$, one end is set to T_w (°C) while the other is maintained at 0 °C, is given by the solution of:

$$\frac{\partial T}{\partial t} = \alpha \cdot \frac{\partial^2 T}{\partial x^2}.$$

Subject to the boundary conditions:

$$T(0, t) = 0; \;\; T(1, t) = T_w, \quad \forall t > 0$$

And to the initial condition: $T(x, 0) = 0, \;\; 0 < x < 1$.
Its analytical solution is given by (Weinberger, 1995):

$$T(x,t) = T_w \cdot x + \sum_{n=1}^{\infty} \frac{2 \cdot T_w \cdot (-1)^n}{\pi \cdot n} \cdot \sin(n \cdot \pi \cdot x) \cdot \exp[-n^2 \cdot \pi^2 \cdot \alpha \cdot t]$$

Example 5.3 The wave equation in one spatial dimension

Determine the displacement $y(x,t)$ of an elastic string fixed at $x = 0$ and $x = L$ if it is initially displaced to the position $y(x,0) = \sin\left(\frac{\pi x}{L}\right)$ and released from rest. No external forces act on the string (Fig. 5.4).

Here, one must solve the wave equation:

$$\frac{\partial^2 y}{\partial t^2} = a^2 \cdot \frac{\partial^2 y}{\partial x^2}.$$

Subject to the boundary conditions $y(0,t) = y(L,t) = 0$, and to the initial conditions:

$$y(x,0) = sen\left(\pi \cdot \frac{x}{L}\right);\ \left.\frac{\partial y}{\partial t}\right|_{t=0} = 0;\ \forall x \in [0, L].$$

The solution to this problem is (Weinberger, 1995):

$$y(x,t) = \sin\left(\frac{\pi \cdot x}{L}\right)\cos\left(a \cdot \pi \cdot \frac{t}{L}\right).$$

The physical phenomena described by the heat conduction equation and the wave equation are different, but both are classified as propagation problems. Therefore, the behavior of the solutions and the methods used to obtain them are different.

Typical examples of propagation include unsteady ideal flow, unsteady mass and heat diffusion, transient boundary-layer flow, transient flow of viscoelastic fluids, and so forth.

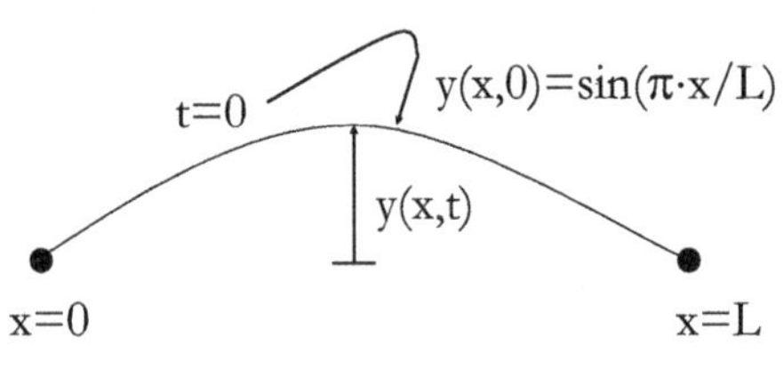

Fig. 5.4 Dynamics of an elastic string

5.1.3 Types of Boundary Conditions

The distinction among different ways of specifying boundary conditions originated in the study of the Laplace equation. Nevertheless, the classification we will now describe applies to any PDE, regardless of its type (elliptic, parabolic, or hyperbolic).[1]

The Dirichlet problem consists of obtaining the solution of the Laplace equation

$$\nabla^2 u = 0 \quad \text{in} \quad D \tag{5.1}$$

in a closed domain D subject to boundary conditions that specify the value of the solution on the boundary B,

$$u = f(s) \quad \text{on} \quad B \tag{5.2}$$

where s is the arclength along B. Boundary condition (5.2) is referred to as a Dirichlet condition.

In contrast, the Neumann problem also requires the solution of (5.1), but now the normal derivative of u on B is specified:

$$\frac{\partial u}{\partial n} = g(s) \quad \text{on} \quad B \tag{5.3}$$

It is also possible to consider that u satisfies more general boundary conditions on B. Thus, one may impose a Robin or mixed boundary condition:

$$a_1(s)\frac{\partial u}{\partial n} + a_2(s)u = h(s) \quad \text{on} \quad B \tag{5.4}$$

The convention of assigning the names Dirichlet, Neumann, or Robin to boundary conditions (5.2), (5.3), and (5.4) has become widespread and is now applied to all types of PDE (linear, nonlinear, equilibrium, propagation, etc.) (Fig. 5.5).

Fig. 5.5 Dirichlet problem

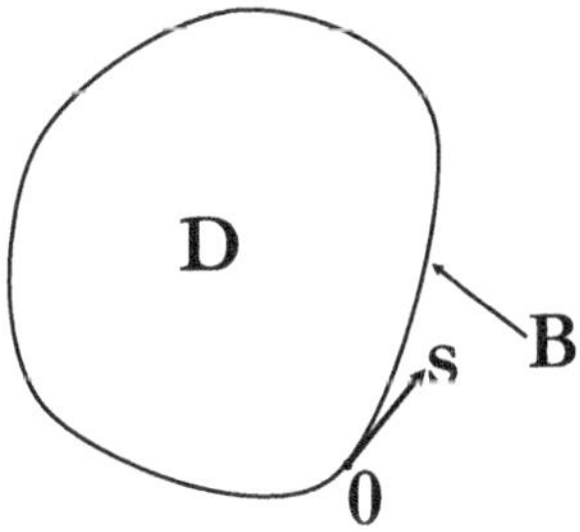

[1] For further mathematical details on the classification of PDEs, see Weinberger (1995).

5.2 The Method of Lines for Propagation Problems with One Spatial Dimension

This method is applicable to propagation problems in which a partial derivative with respect to time appears explicitly for each dependent variable in the problem. If this condition is not satisfied, then the method of lines is not applicable, see Kee and Petzold (1986) for a detailed discussion of this point.

The method replaces the spatial derivatives with finite-difference approximations at all nodes within the domain D; then, by applying the same scheme to the boundary conditions, one obtains a system of ODE whose initial values are specified by the initial condition of the original PDE. The name of the method is because the solution advances in time by computing the value of the function at each node of the spatial grid, thereby constructing the time history of the function values at each node. Next, we present some examples to illustrate how this methodology is applied.

Example 5.4 Application of the method of lines to one-dimensional thermal diffusion

Solve Example 5.2 using the method of lines.

Applying the finite-difference method to a mesh such as that in Fig. 4.4 (with $a = 0$ and $b = 1$), and using the notation introduced there, we have:

$$\frac{dT_i}{dt} = \alpha \cdot \left(\frac{T_{i+1} - 2 \cdot T_i + T_{i-1}}{(\Delta x)^2} \right)$$

$$\rightarrow \frac{dT_i}{d\tau} = T_{i+1} - 2 \cdot T_i + T_{i-1};\, i = 1, 2, \ldots, N-1$$

where a dimensionless time scale has been introduced, $\tau = \alpha t/(\Delta x)^2$. Moreover, from the boundary conditions (Dirichlet in this case), the following additional equations are obtained:

Then, the system of ODE equivalent to the original PDE is given by:

$$T(0,t) = 0 \rightarrow T_0 = 0, t > 0; \quad T(1,t) = T_w \rightarrow T_N = T_w, \quad t > 0;$$

Therefore, the EDO system equivalent to the original EDP is given by:

$$\frac{d}{dt}\begin{bmatrix} T_0 \\ T_1 \\ T_2 \\ \cdot \\ \cdot \\ T_{N-1} \\ T_N \end{bmatrix} = \begin{bmatrix} 0 & 0 & . & . & . & . & 0 \\ 1 & -2 & 1 & . & . & . & 0 \\ 0 & 1 & -2 & 1 & 0 & . & 0 \\ . & . & . & . & . & . & . \\ . & . & . & . & . & . & . \\ 0 & 0 & . & 0 & 1 & -2 & 1 \\ 0 & 0 & . & . & . & . & 0 \end{bmatrix} \cdot \begin{bmatrix} T_0 \\ T_1 \\ T_2 \\ \cdot \\ \cdot \\ T_{N-1} \\ T_N \end{bmatrix}$$

where the initial ODE vector is a vector whose first $N - 1$ components are zero and whose last component is T_w. Figure 5.6 shows the result for the case $N = 40$ and $T_w = 10$ °C. The contour curves correspond to the different dimensionless times τ at which the solution $T(x, \tau)$ has been computed. The solution exhibits the typical features of diffusive problems whose initial condition is a finite temperature gradient: as time advances, the solution is continuous and tends to an equilibrium state, which in this case corresponds to a straight line connecting the points (0, 0) and (1, 10) in the following figure.

Next, we present the Python script that solves the equations with the initial condition that all temperatures are zero except for the value at $x_N = 1$, which is held fixed at $T_N = 10$°C. The system of differential equations is included as a subfunction within the main function.

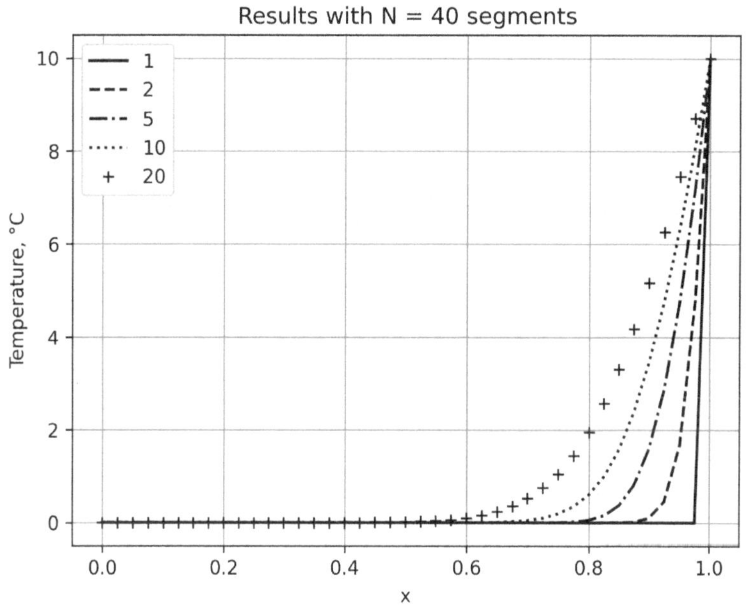

Fig. 5.6 Result of applying the method of lines to Example 5.4

```
import numpy as np
import matplotlib.pyplot as plt
from scipy.integrate import solve_ivp

def slab(N, Twall, tau):
    """
    Function that solves Example 5.4 using the method of lines

    Parameters
    ----------
    N     : number of spatial segments
    Twall : boundary temperature at x = 1
    tau   : array of time points where the solution is evaluated

    Returns
    -------
    y : temperature profiles at the requested time points
    """

    dim = N + 1
    X = np.linspace(0.0, 1.0, dim)

    styles = ['k-', 'k--', 'k-.', 'k:', 'k+', 'ko', 'k*']

    # Initial condition vector
    Y0 = np.zeros(dim)
    Y0[-1] = Twall

    # Solve the ODE system
    sol = solve_ivp(
        fun=lambda t, y: fun(t, y, dim),
        t_span=(tau[0], tau[-1]),
        y0=Y0,
        t_eval=tau,
        method='LSODA'
    )

    y = sol.y.T   # shape: (len(tau), dim)

    # Plot results
    plt.plot(X, y[0, :], styles[0])
    for i in range(1, len(tau)):
        plt.plot(X, y[i, :], styles[i % len(styles)])

    plt.xlabel('x')
    plt.ylabel('Temperature, °C')
    plt.title(
        f'Results with N = {N} segments'
    )
    plt.legend([str(t) for t in tau], loc='best')
    plt.grid(True)
    filename = 'Figure_5_6.png'
    plt.savefig(filename, dpi=300, bbox_inches='tight')
    plt.show()

    return y


def fun(t, y, dim):
    """
    ODE system generated by spatial discretization
    """
    yp = np.zeros(dim)

    for i in range(1, dim - 1):
        yp[i] = y[i - 1] - 2.0 * y[i] + y[i + 1]

    return yp

# generation of figure 5.6
N = 40
Twall = 10
tau = np.array([1, 2, 5, 10, 20])

y = slab(N, Twall, tau)
```

Example 5.5 Application of the method of lines to diffusion and chemical reaction in a porous spherical particle

Consider first-order diffusion and chemical reaction in a spherical, porous solid particle; under transient conditions, the mass balance is expressed by the PDE:

$$\frac{\partial C}{\partial t} = \frac{D_{\text{eff}}}{r^2} \cdot \frac{\partial}{\partial r}\left\{r^2 \cdot \frac{\partial C}{\partial r}\right\} - k \cdot C$$

Subject to the initial and boundary conditions given by:

$$C(r, t = 0) = 0; \quad 0 < r < R;$$

$$\left(\frac{\partial C}{\partial r}\right)(r = 0) = 0; -D_{eff} \cdot \left(\frac{\partial C}{\partial r}\right)(r = 0) = k_m \cdot \left\{C(r = R) - C_f\right\}.$$

The initial conditions indicate that the pellet initially contained no solute and is immersed in a fluid with concentration C_f at $t = 0$. The first boundary condition is due to the symmetry that the concentration must satisfy at $r = 0$, and the second corresponds to continuity of solute flux across the solid–fluid interface at $r = R$.

Using dimensionless variables makes it possible to minimize the number of parameters in the equation to be solved (Aris, 1999); in this case, the following variables are defined:

$$y = \frac{C(r, t)}{C_f}; x = \frac{r}{R}; \tau = \frac{t \cdot D_{eff}}{R^2}$$

With these definitions, the mass-balance equation takes the form:

$$\frac{\partial y}{\partial \tau} = \frac{\partial^2 y}{\partial x^2} + \frac{2}{x} \cdot \frac{\partial y}{\partial x} - \Phi^2 \cdot y; \Phi^2 = \frac{kR^2}{D_{eff}}.$$

Subject to the initial and boundary conditions given by:

$$y(x, 0) = 0; \quad 0 < x < 1;$$

$$\left(\frac{\partial y}{\partial x}\right)(x = 0) = 0; \quad t > 0$$

$$\left(\frac{\partial y}{\partial x}\right)(x = 1) = -Bi_m \cdot \{y(1, t) - 1\}; \ Bi_m = \frac{k_m \cdot R}{D_{eff}}; t > 0$$

Applying the finite-difference method to the preceding dimensionless equation yields the following system of ODE:

$$\frac{dy_i}{d\tau} = \left(\frac{y_{i+1} - 2 \cdot y_i + y_{i-1}}{(\Delta x)^2}\right) + \frac{2}{x_i} \cdot \left(\frac{y_{i+1} - y_{i-1}}{2 \cdot \Delta x}\right) - \Phi^2 \cdot y_i; \quad i = 1, 2, \ldots, N-1$$

The boundary conditions can be written as:

$$\left(\frac{\partial y}{\partial x}\right)(x = 0) = \frac{-3 \cdot y_0 + 4 \cdot y_1 - y_2}{2 \cdot \Delta x} = 0$$

$$\rightarrow \frac{dy_0}{d\tau} = \frac{4}{3} \cdot \frac{dy_1}{d\tau} - \frac{1}{3} \cdot \frac{dy_2}{d\tau}; \quad y_0(0) = 0$$

$$\left(\frac{\partial y}{\partial x}\right)(x = 1) = \frac{3 \cdot y_N - 4 \cdot y_{N-1} + y_{N-2}}{2 \cdot \Delta x} = -Bi_m \cdot (y_N - 1)$$

$$\rightarrow \frac{dy_N}{d\tau} = \frac{4}{(3 + 2 \cdot \Delta x \cdot Bi_m)} \cdot \frac{dy_{N-1}}{d\tau} - \frac{1}{(3 + 2 \cdot \Delta x \cdot Bi_m)} \cdot \frac{dy_{N-2}}{d\tau};$$

$$y_N(0) = \frac{2 \cdot \Delta x \cdot Bi_m}{(3 + 2 \cdot \Delta x \cdot Bi_m)}$$

Then, the numerical solution of the ODE system is obtained together with the initial condition that the solution is zero everywhere except at $x_N = 1$, where it equals $y_N(0)$. Clearly, the solution will depend on the values of the dimensionless Biot number (Bi_m) and the Thiele modulus Φ^2, parameters that are explained next (Finlayson, 2002).

The Biot number can be written as $\mathrm{Bi}_m = \left(R^2/D_{\mathrm{eff}}\right)/(R/k_m)$, that is, the ratio of two characteristic times: diffusion within the porous solid and mass transfer through the fluid film surrounding the pellet. When $\mathrm{Bi}_m \gg 1$, internal diffusive resistance dominates, and one way to mitigate this is to reduce the particle size.

Similarly, the Thiele modulus can be written as $\Phi^2 = \left(R^2/D_{\mathrm{eff}}\right)/(1/k)$, that is, the ratio of a diffusion time to a reaction time. If the reaction is very fast, its characteristic time is small and Φ^2 is large: the concentration at the center of the pellet is small because the solute is consumed before it reaches that location. If diffusion is fast, its characteristic time is small and Φ^2 is small: now the solute concentration cannot vary significantly within the pellet because diffusion maintains the pellet supplied with solute. Therefore, Φ^2 measures the relative importance of diffusion and reaction within the pellet.

Figure 5.7 shows the result of numerical integration using the method of lines for a particular example. As time increases, the concentration profile becomes smoother; for dimensionless times τ greater than 2, the solution no longer changes appreciably.

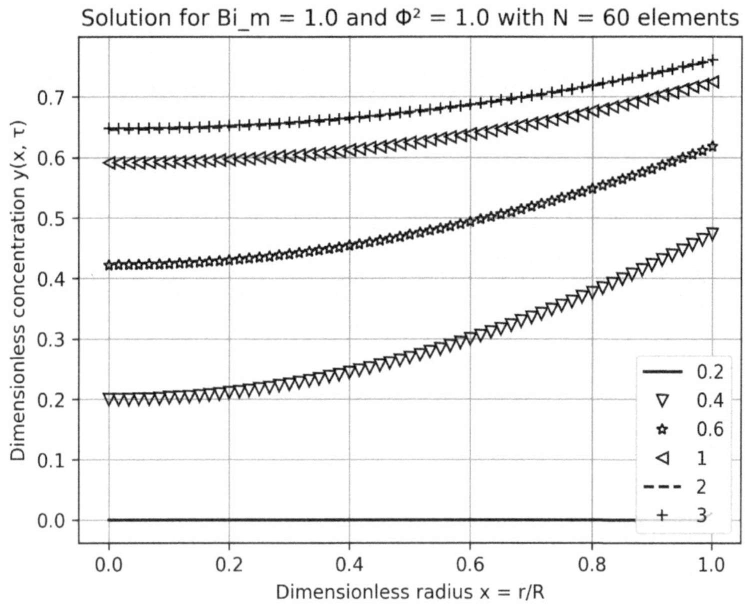

Fig. 5.7 Numerical solution of the equations in Example 5.5

Next, we present the script that solves the ODE system and plots the solution for different integration times.

```
import numpy as np
import matplotlib.pyplot as plt
from scipy.integrate import solve_ivp


def spheric_pellet(n, phi2, Bi, tau):
    """
    Function that computes the time evolution of solute concentration
    in a porous spherical pellet with effective diffusivity D_eff,
    mass transfer coefficient k_m at the fluid side, and radius R.

    Input variables:
    n    : number of nodes in the finite-difference grid
    phi2 : Thiele modulus = k*R^2 / D_eff
    Bi   : Biot number for mass transfer = k_m*R / D_eff
    tau  : vector of dimensionless times where y(x,t) is computed

    Output variables:
    T : vector of integration times (equal to tau)
    Y : concentration values at the N+1 spatial nodes
    """

    # Grid definition
    N = n                          # number of grid segments
```

```
    dx = 1.0 / N                # grid spacing
    X = np.linspace(0.0, 1.0, N + 1)  # spatial domain

    beta = 3.0 + 2.0 * dx * Bi  # numerical factor

    # Initial condition
    Y0 = np.zeros(N + 1)
    Y0[-1] = 2.0 * dx * Bi / beta  # boundary-condition effect in IC

    # Time integration (ode113 equivalent)
    sol = solve_ivp(
        fun=lambda t, y: system(t, y, N, phi2, beta, X, dx),
        t_span=(tau[0], tau[-1]),
        y0=Y0,
        t_eval=tau,
        method="LSODA"
    )

    T = sol.t
    Y = sol.y.T   # shape: (len(tau), N+1)

    # Plot results
    styles = ['k-', 'k--', 'kd', 'k*', 'ko', 'k+']

    plt.plot(X, Y[0, :], styles[0])
    for i in range(1, len(T)):
        plt.plot(X, Y[i, :], styles[i % len(styles)])

    plt.title(
        f"Solution for Bi_m = {Bi} and Φ² = {phi2} with N = {N} elements"
    )
    plt.xlabel("Dimensionless radius x = r/R")
    plt.ylabel("Dimensionless concentration y(x, τ)")
    plt.legend([f"{t:.3g}" for t in T], loc="best")
    plt.grid(True)
    plt.show()

    return T, Y


def system(t, y, N, phi2, beta, X, dx):
    """
    Sub-function that defines the system of differential equations
    for the method of lines
    """

    yp = np.zeros(N + 1)

    for i in range(1, N):
        yp[i] = (
            (y[i - 1] - 2.0 * y[i] + y[i + 1]) / dx**2
            + (y[i + 1] - y[i - 1]) / (X[i] * dx)
            - phi2 * y[i]
        )

    # Boundary conditions
    yp[0] = (4.0 * yp[1] - yp[2]) / 3.0
    yp[N] = (4.0 * yp[N - 1] - yp[N - 2]) / beta

    return yp

# generation of figure 5.7
n = 60
phi2 = 1.0
Bi = 1.0
tau = np.array([0.2, 0.4, 0.6, 1.0, 2.0, 3.0])

T, Y = spheric_pellet(n, phi2, Bi, tau)
```

5.3 The Finite-Difference Method for Equilibrium Problems

For equilibrium problems in more than one spatial dimension, the computational methodology is the same as in the one-dimensional case (Chap. 4), except that additional indices are required to denote the finite-difference approximations, since the grid is now 2D or 3D.

Example 5.6 Application of the finite-difference method to heat conduction in two spatial dimensions

Consider the steady-state 2D heat-conduction problem defined in Example 5.1. Let us construct a mesh in the domain [0, 1]⊗[0, 1], characterized by the expressions:

$$x_i = i \cdot \Delta x;\ \Delta x = \frac{1}{N};\quad i = 0, 1, 2, \ldots, N$$

$$y_j = j \cdot \Delta y;\ \Delta y = \frac{1}{M};\quad j = 0, 1, 2, \ldots, M$$

Figure 5.8 shows an example of a 2D mesh in which the number of nodes is, in general, different in the x and y coordinates (in this case, $N = 8$ and $M = 4$, $\Delta x = \Delta y = 1$).

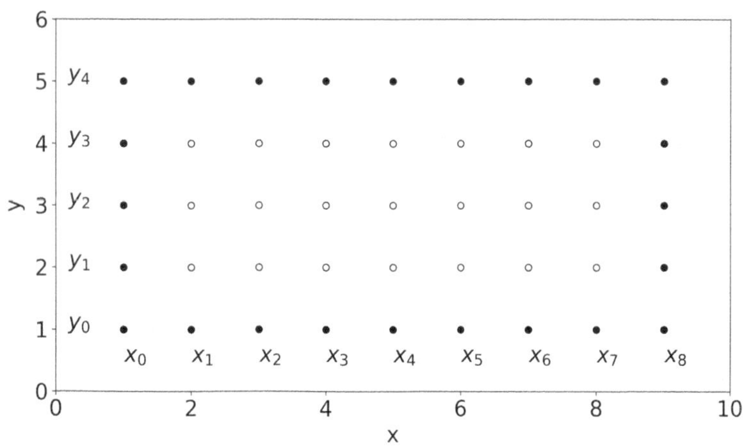

Fig. 5.8 2D mesh for solving the Laplace equation. The filled black circles stand for boundary nodes

Using finite differences, Eq. (5.1) is expressed as follows:

$$\nabla^2 T(x_i, y_j) = \frac{T_{i-1,j} - 2T_{i,j} + T_{i+1,j}}{(\Delta x)^2} + \frac{T_{i,j-1} - 2T_{i,j} + T_{i,j+1}}{(\Delta y)^2} = 0;$$
$$i = 1, 2, ..., N - 1; j = 1, 2, \ldots, M - 1$$

And the boundary conditions are as follows:

$$T_{i,0} = T_w; T_{i,M} = 0; \quad i = 0, 1, 2, \ldots, N;$$
$$T_{0,j} = 0; T_{N,j} = 0; \quad j = 1, 2, ..., M - 1.$$

The discretized Laplace equation can be written as the following system of linear equations:

$$-T_{i-1,j} + 2 \cdot (1 + \beta^2) \cdot T_{i,j} - T_{i+1,j} = \beta^2 \cdot (T_{i,j-1} + T_{i,j+1}); \beta = \Delta x / \Delta y$$
$$i = 1, 2, \ldots, N - 1; j = 1, 2, \ldots, M - 1$$

This can be viewed as an iterative formula to successively compute the temperatures of the rows $j = 1, 2, \ldots, M - 1$. The equation is solved only for the interior nodes ($i = 1, 2, \ldots, N - 1$), since the values at the boundaries do not change (they are steady boundary conditions). Starting from row $j = 1$ onward, the row sweep completes one full iteration cycle upon reaching row $M - 1$. To solve the preceding system of equations, we use Moler efficient tridisolve algorithm (Moler, 2008), in which the three diagonals of the matrix are always the same vectors throughout the iteration (see Example 1.2). The right-hand side of the preceding system is computed using the result from iteration k to evaluate $T_{i,j+1}$(the value in the next row) and using $T_{i,j-1}$ from iteration $k + 1$ in the previously computed row. In this way, the available

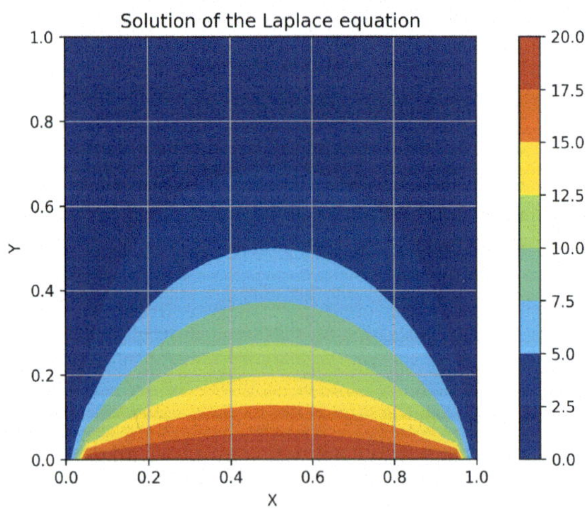

Fig. 5.9 Numerical solution of Example 5.6 with $N = M = 20$

values are updated immediately for the right-hand-side vector, thereby accelerating convergence.

Figure 5.9 shows convergent results for the case $T_w = 20°C$. For a grid with $N = M = 20$, the method requires 109 iterations to reduce the relative error $\text{norm}(T^{k+1} - T^k)/\text{norm}(T^{k+1})$ below 10^{-4}. Upon verifying the solution, the maximum residual in the system of equations is found to be 0.00156.

Below is the script that solves the example.

```
import numpy as np
import matplotlib.pyplot as plt


def tridisolve(a, b, c, d):
    """
    Solves a tridiagonal linear system using the Thomas algorithm

    a : subdiagonal (length n-1)
    b : main diagonal (length n)
    c : superdiagonal (length n-1)
    d : right-hand side vector (length n)
    """
    n = len(b)
    cp = np.zeros(n-1)
    dp = np.zeros(n)

    cp[0] = c[0] / b[0]
    dp[0] = d[0] / b[0]

    for i in range(1, n-1):
        denom = b[i] - a[i-1] * cp[i-1]
        cp[i] = c[i] / denom
        dp[i] = (d[i] - a[i-1] * dp[i-1]) / denom

    dp[n-1] = (d[n-1] - a[n-2] * dp[n-2]) / (b[n-1] - a[n-2] * cp[n-2])

    x = np.zeros(n)
    x[-1] = dp[-1]
    for i in reversed(range(n-1)):
        x[i] = dp[i] - cp[i] * x[i+1]

    return x


def laplace(N, M, Tw, errmax=1e-3, maxit=100):
    """
    Function that solves the Laplace equation of Example 5.1
    using finite-difference methods

    N : number of nodes in the x direction, x in [0, 1]
    M : number of nodes in the y direction, y in [0, 1]
    Tw : temperature on the lower boundary of the square [0,1] x [0,1]
    errmax : maximum allowed error
    maxit : maximum number of iterations allowed

    Returns
    -------
    T : (N+1) x (M+1) matrix with temperature values
    err : final maximum error
    iter : number of iterations performed
    lapmax : maximum value of the discrete Laplacian (diagnostic)
```

```
    """

    # Common parameters and initial temperature field
    dx = 1.0 / N
    dy = 1.0 / M
    beta2 = (dx / dy) ** 2

    a = -np.ones(N - 1)
    b = 2.0 * (1.0 + beta2) * np.ones(N - 1)
    c = -np.ones(N - 1)
    r = np.zeros(N - 1)

    T = np.zeros((N + 1, M + 1))

    # Boundary conditions
    T[:, 0] = Tw          # bottom boundary
    T[:, -1] = 0.0        # top boundary
    T[0, :] = 0.0         # left boundary
    T[-1, :] = 0.0        # right boundary

    # Iteration loop
    err = 1.0
    iter = 0

    while err > errmax and iter < maxit:
        Told = T.copy()

        # Row-sweep loop
        for j in range(1, M):
            for i in range(N - 1):
                r[i] = beta2 * (T[i + 1, j - 1] + Told[i + 1, j + 1])

            v = tridisolve(a, b, c, r)
            T[1:N, j] = v

        err = np.linalg.norm(T - Told) / np.linalg.norm(T)
        iter += 1

    # Maximum discrete Laplacian (optional diagnostic)
    lapmax = np.max(np.abs(T[2:, 1:-1] + T[:-2, 1:-1] +
                           T[1:-1, 2:] + T[1:-1, :-2] -
                           4 * T[1:-1, 1:-1]))

    # Results visualization
    x = np.linspace(0.0, 1.0, N + 1)
    y = np.linspace(0.0, 1.0, M + 1)

    plt.contourf(x, y, T.T, cmap="jet")
    plt.colorbar()
    plt.title("Solution of the Laplace equation")
    plt.xlabel("X")
    plt.ylabel("Y")
    plt.axis("square")
    plt.grid(True)
    plt.show()

    return T, err, iter, lapmax

T, err, iters, lapmax = laplace(
    N=20,
    M=20,
    Tw=20.0,
    errmax=1e-4,
    maxit=200
)
print(iters)
print(lapmax)
```

Note that the tridiagonal matrix is dominant, so the over-relaxation method can also be used to solve the temperatures in each row of the mesh, introducing a relaxation parameter ω to accelerate convergence. The reader is encouraged to modify the previous code to include over-relaxation and to compare the number of iterations needed to reach the same convergence criterion as in the example. For the case of the Laplace equation, it has been shown that the optimal value of the parameter ω is given by (Anderson et al., 2011):

$$\omega_{opt} = \frac{2}{1+\sqrt{1-\sigma^2}};\ \sigma = \frac{1}{1+\beta^2}\left\{\cos\left(\frac{\pi}{N}\right) + \beta^2 \cdot \cos\left(\frac{\pi}{M}\right)\right\} \tag{5.5}$$

In the most general case, when Laplace's equation includes on the right-hand side an energy generation term of the form $g(x, y)$, the equation becomes the discrete Poisson equation:

$$\nabla^2 T\left(x_i, y_j\right) = \frac{T_{i-1,j} - 2T_{i,j} + T_{i+1,j}}{(\Delta x)^2} + \frac{T_{i,j-1} - 2T_{i,j} + T_{i,j+1}}{(\Delta y)^2} = g\left(x_i, y_j\right);$$
$$i = 1, 2, \ldots, N-1; j = 1, 2, \ldots, M-1 \tag{5.6}$$

The resulting system of equations is now nonlinear, and one must apply the techniques seen in Chap. 2—such as Newton method—to solve the equations efficiently. For example, one can solve row by row or column by column to exploit the structure of the equations, since the Jacobian has a simple structure. Moreover, because the solution is expected to be continuous, the solution found for the most recently computed column (or row) can be used as the initial guess for the nonlinear system, starting with the boundary condition in the first case. For more details, see Anderson et al. (2011) and Sewell (1988).

5.4 Finite-Difference Methods in Propagation Problems with One Spatial Dimension

Finite-difference methods can be applied to propagation problems by introducing finite differences in both the spatial and temporal coordinates.

Consider solving Example 5.2 using finite differences (in space and time). We use a time discretization so that $t_j = j \cdot \Delta t, j = 0, 1, \ldots, M$, and, using a superscript to denote the time step, the equations to be solved are:

$$\frac{dT_i^j}{d\tau} = T_{i-1}^j - 2 \cdot T_i^j + T_{i+1}^j; \quad i = 1, 2, \ldots, N-1; \quad j = 0, 1, \ldots, M \tag{5.7}$$

Subject to the boundary conditions:

$$T(0,t) = 0 \to T_0^j = 0; T(1,t) = T_w \to T_N^j = T_w; \quad j = 0, 1, \ldots, M \tag{5.7a}$$

and to the initial condition:

$$T(x,0) = 0 \to T_i^0 = 0; \quad i = 0, 1, \ldots, N \tag{5.7b}$$

Now, the time derivative can be represented by finite differences in several ways, some of which we detail below.

(a) Forward differences in time.

$$\begin{aligned} \frac{dT_i^j}{d\tau} &= \frac{T_i^{j+1} - T_i^j}{\Delta\tau} = T_{i-1}^j - 2 \cdot T_i^j + T_{i+1}^j; \\ \to T_i^{j+1} &= \Delta\tau \cdot \left(T_{i-1}^j + T_{i+1}^j\right) + (1 - 2 \cdot \Delta\tau) \cdot T_i^j \end{aligned} \tag{5.8}$$

In this case, the time derivative is approximated using forward differences, yielding an explicit scheme. This equation is valid for propagating the temperature values at the internal nodes ($i = 1, 2, \ldots, N-1$); the values at $i = 0$ and $i = N$ are computed only once using the boundary conditions (5.7a). Equation (5.8) represents a numerical integration algorithm for which stability must be guaranteed. It can be shown that the stability condition for the explicit method in Eq. (5.8) is given by[2]:

$$\tau = \alpha \cdot \frac{\Delta t}{(\Delta x)^2} \le \frac{1}{2} \to \Delta t \le \frac{(\Delta x)^2}{2 \cdot \alpha} \tag{5.9}$$

(b) Backward differences in time.

In this case, Eq. (5.8) takes the form of an implicit scheme:

$$\begin{aligned} \frac{dT_i^{j+1}}{d\tau} &= \frac{T_i^{j+1} - T_i^j}{\Delta\tau} = T_{i-1}^{j+1} - 2 \cdot T_i^{j+1} + T_{i+1}^{j+1}; \\ \to \Delta\tau \cdot T_{i-1}^{j+1} &- (1 + 2 \cdot \Delta\tau) \cdot T_i^{j+1} + \Delta\tau \cdot T_{i+1}^{j+1} = -T_i^j \end{aligned} \tag{5.10}$$

[2] See for example, Rice and Do (2012), Chapter 12.

This yields a tridiagonal system of equations to update the interior temperature nodes, which can be solved easily using the Thomas algorithm (Chap. 1) already implemented in Example 1.2. It can be shown that this method is stable for any value of the integration step $\Delta\tau$. However, the method has low accuracy in time (first order) while remaining second order in x due to the use of second-order finite differences in the spatial direction. Next, we discuss how to improve the accuracy of numerical integration in time.

(c) Improving accuracy in time.

Consider the midpoint of the time integration interval: $t_{j+1/2} = t_j + \Delta t/2$. Applying finite differences in time at this midpoint, one can show that:

$$\left(\frac{\partial T}{\partial t}\right)\left(x_i, t = t_{j+1/2}\right) = \frac{T_i^{j+1} - T_i^j}{\Delta\tau} + O\left((\Delta\tau)^2\right) \tag{5.11}$$

and approximating the right-hand side of the original PDE using the trapezoidal rule at times t_j and t_{j+1}, we obtain the scheme of Crank–Nicolson (1947):

$$\begin{aligned} &\frac{T_i^{j+1} - T_i^j}{\Delta\tau} = \frac{1}{2}\cdot\left\{\left(T_{i-1}^j - 2\cdot T_i^j + T_{i+1}^j\right) + \left(T_{i-1}^{j+1} - 2\cdot T_i^{j+1} + T_{i+1}^{j+1}\right)\right\} \\ &\to T_{i-1}^{j+1} - 2\cdot(1 + 1/\Delta\tau)\cdot T_i^{j+1} + T_{i+1}^{j+1} = -T_{i-1}^j + 2\cdot(1 - 1/\Delta\tau)\cdot T_i^j - T_{i+1}^j. \end{aligned} \tag{5.12}$$

This again yields a tridiagonal system of equations to advance the solution at the interior nodes, but this time the integration method has accuracy $O\left((\Delta\tau)^2\right)$. This formulation is known as the Crank–Nicolson method and is stable in both the spatial and temporal dimensions. Moreover, the approximation is more accurate in time and therefore allows taking larger time steps, as seen in the following example.

Example 5.7 Application of the finite-difference method to transient heat conduction in one spatial dimension

Figure 5.10 shows the results of integrating the differential equations according to the three methods described in this Section. A spatial grid with $\Delta x = 0.1$ was used for the integration, and three different values of $\Delta\tau$ (0.1, 0.25, and 0.5) were employed. The solution at $x = 0.9$ is plotted for $\tau = 0.5$, 1, 2, 3, and 4.

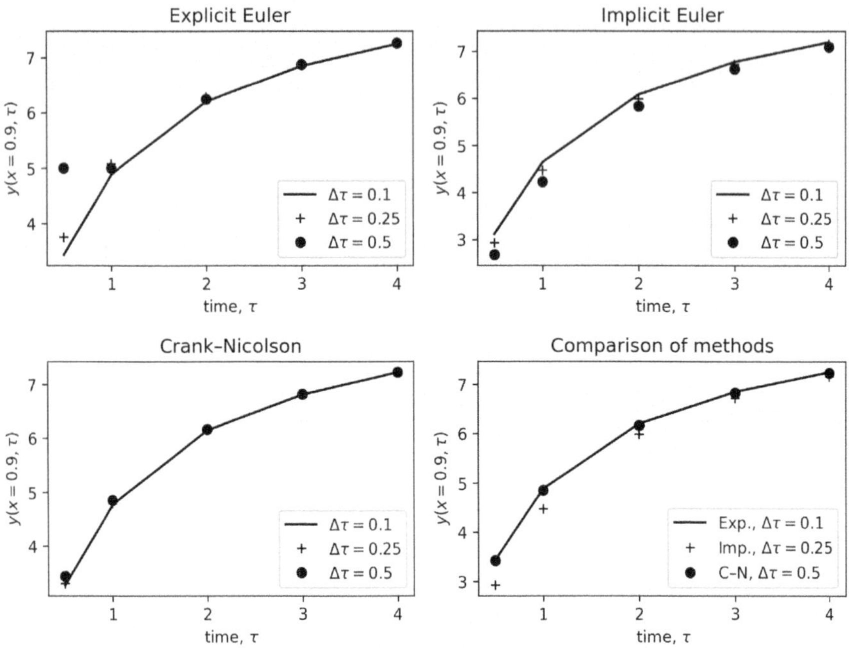

Fig. 5.10 Comparison of results with different methods applied to Example 5.7

Figure 5.10 shows that the Crank–Nicolson method is accurate in time, as its convergence is rapid; in contrast, the explicit Euler method is the slowest to converge, and the implicit Euler method has intermediate convergence because, although it is stable, it is less accurate than the Crank–Nicolson method. With $\Delta\tau = 0.5$, the Crank–Nicolson method yields better results than implicit Euler with $\Delta\tau = 0.25$ (with roughly double the computational effort) and than explicit Euler with $\Delta\tau = 0.1$ (with roughly five times the computational effort).

Below we present the script that solves the equations of the example; as an exercise, the reader is left to program the script that produces Fig. 5.10.

```
import numpy as np


def Ej_5_7(N, Tw):
    """
    Example 5.7

    Input variables:
    N  : number of spatial grid elements
    Tw : wall temperature at x = 1

    Output variables:
    Output1 : Explicit Euler results (Eq. 5.33), shape (3,5)
    Output2 : Implicit Euler results (Eq. 5.35), shape (3,5)
    Output3 : Crank-Nicolson results (Eq. 5.37), shape (3,5)
    """

    deltasT = np.array([0.1, 0.25, 0.5])       # time integration steps
    taus = np.array([0.5, 1, 2, 3, 4])         # integration times

    Output1 = np.zeros((3, 5))
    Output2 = np.zeros((3, 5))
    Output3 = np.zeros((3, 5))

    for i in range(3):
        dt = deltasT[i]

        # Constant vectors defining the tridiagonal systems
        a = dt * np.ones(N - 1)
        b = -(1 + 2 * dt) * np.ones(N - 1)
        c = dt * np.ones(N - 1)

        d = np.ones(N - 1)
        e = -2 * (1 + 1 / dt) * d
        f = d

        # ---------- Explicit Euler method ----------
        t0 = 0.0
        y0 = np.zeros(N + 1)
        y0[-1] = Tw

        for j in range(5):
            tf = taus[j]
            Tout = fun1(N, dt, t0, tf, y0)
            Output1[i, j] = Tout[N - 1]
            y0 = Tout
            t0 = tf

        # ---------- Implicit Euler method ----------
        t0 = 0.0
        y0 = np.zeros(N + 1)
        y0[-1] = Tw

        for j in range(5):
            tf = taus[j]
            Tout = fun2(N, dt, t0, tf, y0, Tw, a.copy(), b.copy(), c.copy())
            Output2[i, j] = Tout[N - 1]
            y0 = Tout
            t0 = tf

        # ---------- Crank-Nicolson method ----------
        t0 = 0.0
        y0 = np.zeros(N + 1)
        y0[-1] = Tw
```

```
        for j in range(5):
            tf = taus[j]
            Tout = fun3(N, dt, t0, tf, y0, Tw, d.copy(), e.copy(), f.copy())
            Output3[i, j] = Tout[N - 1]
            y0 = Tout
            t0 = tf

    return Output1, Output2, Output3

def fun1(N, dt, t0, tf, y0):
    """Explicit Euler method"""
    y = y0.copy()
    M = int((tf - t0) / dt)

    for _ in range(M):
        y_new = y.copy()
        for i in range(1, N):
            y_new[i] = dt * (y[i - 1] + y[i + 1]) + (1 - 2 * dt) * y[i]
        y = y_new

    return y

def fun2(N, dt, t0, tf, y0, Tw, a, b, c):
    """Implicit Euler (Backward Euler) method"""
    y = y0.copy()
    M = int((tf - t0) / dt)

    for _ in range(M):
        r = -y[1:N]
        r[-1] -= dt * Tw
        v = tridisolve(a, b, c, r)
        y = np.concatenate(([y[0]], v, [y[-1]]))

    return y

def fun3(N, dt, t0, tf, y0, Tw, d, e, f):
    """Crank-Nicolson method"""
    y = y0.copy()
    M = int((tf - t0) / dt)

    for _ in range(M):
        r = np.zeros(N - 1)
        for i in range(N - 1):
            r[i] = -y[i] + 2 * (1 - 1 / dt) * y[i + 1] - y[i + 2]
        r[-1] -= Tw

        v = tridisolve(d, e, f, r)
        y = np.concatenate(([y[0]], v, [y[-1]]))

    return y

def tridisolve(a, b, c, d):
    """
    Solves a tridiagonal system using the Thomas algorithm
    """
    n = len(d)
    b = b.copy()
    x = d.copy()

    for j in range(n - 1):
        mu = a[j] / b[j]
        b[j + 1] -= mu * c[j]
        x[j + 1] -= mu * x[j]

    x[-1] /= b[-1]
    for j in range(n - 2, -1, -1):
        x[j] = (x[j] - c[j] * x[j + 1]) / b[j]

    return x
```

5.5 Solving Propagation Problems in Two Spatial Dimensions

Let us consider the two-dimensional unsteady heat conduction equation:

$$\frac{\partial T}{\partial t} = \alpha \left\{ \frac{\partial^2 T}{\partial x^2} + \frac{\partial^2 T}{\partial y^2} \right\} = 0; \quad 0 < x < 1; \quad 0 < y < 1; \quad t > 0. \tag{5.13}$$

When finite differences are applied to obtain a discrete representation of the two-dimensional Laplacian and of the left-hand-side time derivative, the following general expression is obtained (using time as a superscript):

$$\frac{T_{i,j}^{k+1} - T_{i,j}^{k}}{\Delta t} = \alpha \nabla^2 T(x_i, y_j). \tag{5.14}$$

Again, we have three options to advance the solution of Eq. (5.13) in time:

- Explicit method: $T_{i,j}^{k+1} = T_{i,j}^{k} + \alpha \Delta t \cdot \nabla^2 T(x_i, y_j, t_k)$
- Implicit method: $T_{i,j}^{k+1} = T_{i,j}^{k} + \alpha \Delta t \cdot \nabla^2 T(x_i, y_j, t_{k+1})$
- Crank–Nicholson: $T_{i,j}^{k+1} = T_{i,j}^{k} + \frac{\alpha \Delta t}{2} \cdot \{\nabla^2 T(x_i, y_j, t_k) + \nabla^2 T(x_i, y_j, t_{k+1})\}$

where the third argument in the discretized Laplacian represents the time at which it is evaluated. It can be shown that the respective truncation errors of each of the above approximations are given by:

- Explicit method: $O[\Delta t, (\Delta x)^2, (\Delta y)^2]$
- Implicit method: $O[\Delta t, (\Delta x)^2, (\Delta y)^2]$
- Crank–Nicholson: $O[(\Delta t)^2, (\Delta x)^2, (\Delta y)^2]$

Regarding stability conditions, the explicit method requires that the following condition be always satisfied:

$$r = \alpha \cdot \Delta t \cdot \left\{ \frac{1}{(\Delta x)^2} + \frac{1}{(\Delta y)^2} \right\} < \frac{1}{2} \tag{5.15}$$

Then, for a fixed value of the diffusivity α, Eq. (5.15) requires decreasing the integration step as the grid is refined; that is, if higher spatial accuracy is required, the restriction on the time step becomes increasingly severe. In contrast, the other two methods are unconditionally stable for any value of Δt.

However, the linear equations that define the Laplacian in terms of the temperature values at the nodes (i,j) no longer have a tridiagonal structure but rather correspond to a sparse matrix. Therefore, the advantage of a tridiagonal matrix, as in the one-dimensional case, is lost. This makes the Crank–Nicolson method less attractive in the two-dimensional case.

We now present two variants that allow Eq. (5.13) to be solved more efficiently.

5.5.1 The Alternating Direction Implicit Method (ADI)

Let us now consider Eq. (5.13), advancing the solution over two equal time subintervals $\Delta t/2$, as follows (Anderson et al., 2011):

(a) We advance by a time $\Delta t/2$, and in the Laplacian, we evaluate the derivatives with respect to x and y at times $t^{k+1/2}$ and t^k, respectively:

$$\frac{T_{i,j}^{k+1/2}-T_{i,j}^{k}}{\alpha(\Delta t/2)}=\frac{1}{(\Delta x)^2}\left\{T_{i-1,j}^{k+1/2}-2T_{i,j}^{k+\frac{1}{2}}+T_{i+1,j}^{k+1/2}\right\}+\frac{1}{(\Delta y)^2}\left\{T_{i,j-1}^{k}-2T_{i,j}^{k}+T_{i,j+1}^{k}\right\}$$

Defining $\beta=\frac{\Delta x}{\Delta y}$, $\gamma=\frac{2(\Delta x)^2}{\alpha\Delta t}$, the previous equations become:

$$T_{i-1,j}^{k+1/2}-(2+\gamma)T_{i,j}^{k+\frac{1}{2}}+T_{i+1,j}^{k+\frac{1}{2}}=-\beta^2\left\{T_{i,j-1}^{k}+T_{i,j+1}^{k}\right\}+\left(2\beta^2-\gamma\right)T_{i,j}^{k}\quad i=1,2,\ldots,N-1 \tag{5.16}$$

where the right-hand side is known, and therefore the above system of equations can be solved column by column using the tridiagonal algorithm. This procedure yields a first update of the solution up to time $t^{k+1/2}$.

(b) We now advance by a time $\Delta t/2$, and in the Laplacian we evaluate the derivatives with respect to x and y at times $t^{k+1/2}$ and t^{k+1}, respectively, thus generating an update in the alternating direction:

$$\frac{T_{i,j}^{k+1}-T_{i,j}^{k+1/2}}{\alpha(\Delta t/2)}=\frac{1}{(\Delta x)^2}\left\{T_{i-1,j}^{k+1/2}-2T_{i,j}^{k+\frac{1}{2}}+T_{i+1,j}^{k+1/2}\right\}+\frac{1}{(\Delta y)^2}\left\{T_{i,j-1}^{k+1}-2T_{i,j}^{k+1}+T_{i,j+1}^{k+1}\right\}.$$

This produces the tridiagonal system of equations:

$$\beta^2T_{i,j-1}^{k+1}-\left(2\beta^2+\gamma\right)T_{i,j}^{k+1}+\beta^2T_{i,j+1}^{k+1}=-\left\{T_{i-1,j}^{k+\frac{1}{2}}+T_{i+1,j}^{k+\frac{1}{2}}\right\}+(2-\gamma)T_{i,j}^{k+\frac{1}{2}}\quad i=1,2,\ldots,N-1 \tag{5.17}$$

The sequential combination of Eqs. (5.16) and (5.17) correspond to the ADI method, which is unconditionally stable for any value of Δt. This method has second-order truncation errors in both time and space, which allows it to be applied efficiently to the solution of transient problems in 2D.

Example 5.8 Application of the ADI method to the development of laminar flow in a square-duct cross section.

Consider a duct with a square cross section of dimensions $L = H = 3$ cm. The fluid (density $\rho = 1$ g/cm^3, viscosity $\mu = 0.03$ g/(cm·s)) is initially at rest, and an axial pressure gradient $\partial p/\partial z = -0.7$ dyn/ cm^3 is applied. Compute the velocity distribution at times $t = 0, 10, \ldots, 100$ and verify whether a steady axial velocity profile has been reached in the duct.

The momentum balance in the axial direction can be written as:

$$\frac{\partial v_Z}{\partial t} = -\frac{1}{\rho}\frac{\partial p}{\partial z} + \frac{\mu}{\rho}\left\{\frac{\partial^2 v_Z}{\partial x^2} + \frac{\partial^2 v_Z}{\partial y^2}\right\}$$

Applying the ADI method equations to the previous equation using a grid with $\beta = 1$, the following expressions are obtained to determine the spatiotemporal distribution of the axial velocity V:

$$\begin{aligned} &V_{i-1,j}^{k+1/2} - (2+\gamma)V_{i,j}^{k+\frac{1}{2}} + V_{i+1,j}^{k+\frac{1}{2}} \\ &= g - \left\{V_{i,j-1}^{k} + V_{i,j+1}^{k}\right\} + (2-\gamma)V_{i,j}^{k} \quad i = 1, 2, \ldots, N-1 \end{aligned}$$

$$\begin{aligned} &V_{i,j-1}^{k+1} - (2+\gamma)V_{i,j}^{k+1} + V_{i,j+1}^{k+1} \\ &= g - \left\{V_{i-1,j}^{k+\frac{1}{2}} + V_{i+1,j}^{k+\frac{1}{2}}\right\} + (2-\gamma)V_{i,j}^{k+\frac{1}{2}} \quad i = 1, 2, \ldots, N-1 \end{aligned}$$

where $g = (\Delta x)^2/\mu \cdot (\partial p/\partial z)$. Figure 5.11 shows the 2D profile of the axial velocity and the evolution of the velocity at the geometric center of the duct, indicating that the steady-state condition has been reached at $t = 100$ s. The script is also provided.

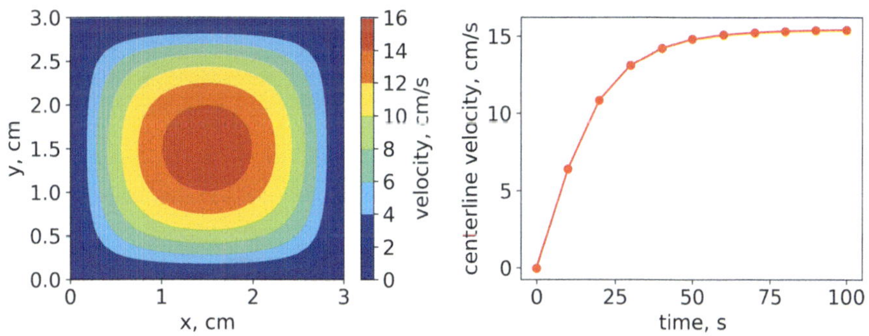

Fig. 5.11 Solution of Example 5.8 using the ADI method

```python
import numpy as np
import matplotlib.pyplot as plt

def Flow_ADI(L, H, N, M, K, dpdz, rho, mu, tf):
    """
    Solves the transient flow in a rectangular duct using the ADI method.

    Parameters
    ----------
    L, H : float
        Length and height of the cross section (cm)
    N, M, K : int
        Number of discrete segments in x, y, and time directions
    dpdz : float
        Axial pressure gradient (dyn/cm^3)
    rho, mu : float
        Fluid density (g/cm^3) and viscosity (g/cm/s)
    tf : float
        Final integration time (s)

    Returns
    -------
    V : ndarray
        Velocity distributions at different times (third dimension)
    VC : ndarray
        Centerline velocity as a function of time (length K+1)
    """

    # Numerical model parameters
    dx = L / N
    dy = H / M
    dt = tf / K

    gamma = 2 * dx**2 / (mu * dt / rho)
    g = dx**2 * dpdz / mu

    # Velocity arrays
    V = np.zeros((N + 1, M + 1, K + 1))
    Vint = np.zeros((N + 1, M + 1))

    # Tridiagonal matrix for implicit solves
    A = np.zeros((N + 1, N + 1))
    b = np.zeros(N + 1)

    for i in range(1, N):
        A[i, i - 1:i + 2] = [1, -(2 + gamma), 1]

    A[0, 0] = 1
    A[N, N] = 1

    # ADI (IDA) time-stepping loop
    for k in range(K):

        # First half-step (implicit in x)
        for j in range(1, M):
            for i in range(1, N):
                b[i] = (
                    g
                    - V[i, j - 1, k]
                    - V[i, j + 1, k]
                    + (2 - gamma) * V[i, j, k]
                )
            v = np.linalg.solve(A, b)
            Vint[:, j] = v

        # Second half-step (implicit in y)
        for i in range(1, N):
            for j in range(1, M):
                b[j] = (
                    g
```

```
                  - Vint[i - 1, j]
                  - Vint[i + 1, j]
                  + (2 - gamma) * Vint[i, j]
              )
          v = np.linalg.solve(A, b)
          V[i, :, k + 1] = v

  # Coordinates
  x = np.linspace(0, L, N + 1)
  y = np.linspace(0, H, M + 1)
  T = np.linspace(0, tf, K + 1)

  # Plot results
  plt.figure(figsize=(10, 4))

  plt.subplot(1, 2, 1)
  plt.contourf(x, y, V[:, :, -1].T)
  plt.xlabel('x, cm')
  plt.ylabel('y, cm')
  plt.colorbar(label='velocity, cm/s')

  # Centerline velocity
  i1 = N // 2
  i2 = M // 2
  VC = V[i1, i2, :]

  plt.subplot(1, 2, 2)
  plt.plot(T, VC, '-ro')
  plt.xlabel('time, s')
  plt.ylabel('centerline velocity, cm/s')

  plt.tight_layout()
  plt.show()

  return V, VC
```

5.5.2 The Method of Lines in Two Spatial Dimensions

If we approximate Eq. (5.13) using finite differences on the right-hand side and interpret the partial derivative on the left-hand side at the point with coordinates (x_i, y_j), we obtain:

$$\frac{\mathrm{d}}{\mathrm{d}t}T_{i,j} = \frac{1}{(\Delta x)^2}\{T_{i-1,j} - 2T_{i,j} + T_{i+1,j}\} + \frac{1}{(\Delta y)^2}\{T_{i,j-1} - 2T_{i,j} + T_{i,j+1}\}; \quad i = 1, \ldots, N-1; j = 1, \ldots, M-1 \tag{5.18}$$

This yields a system of $(N-1) \times (M-1)$ ODE for the interior grid nodes. For the boundary conditions, the corresponding ODE that apply at those boundaries must be generated. For example, if a zero heat-flux condition is imposed at $y = y_{\max}$, then the relevant equation is:

$$T_{i,M-2} - 4T_{i,M-1} + 3T_{i,M} = 0 \rightarrow \frac{\mathrm{d}}{\mathrm{d}t}T_{i,M}$$

$$= \frac{4}{3}\frac{\mathrm{d}}{\mathrm{d}t}T_{i,M-1} - \frac{1}{3}\frac{\mathrm{d}}{\mathrm{d}t}T_{i,M-2};\, i = 0, 1, \ldots, N \tag{5.18a}$$

In this way, a system of $(N + 1) \times (M + 1)$ ODE is obtained, whose initial conditions follow the initial condition for $T(x, y, t = 0)$. The advantage of this formulation, relative to the ADI method, is that numerical integration can handle nonlinear terms, such as heat generation due to chemical reaction within the domain, or boundary conditions involving radiation or natural convection. The disadvantage is that the dimension of the ODE system increases rapidly as the grid is refined, and the system becomes increasingly stiff. If higher solution accuracy is required, one alternative is to switch to the explicit Euler method and reduce the integration step until numerical stability is achieved. The other option is to solve Eq. (5.13) using orthogonal collocation methods, which yield lower-dimensional ODE systems (see Rice and Do, Chapter 8, for a comprehensive presentation of that methodology).

To implement the method of lines using Python numerical integrators, Eq. (5.18) must be written by arranging all derivatives in a column vector, for which the reshape function is used. The stages of the algorithm are as follows:

i. The initial matrix $T_0(i, j)$ is generated, which must satisfy the ICs and BCs.
ii. Reshape is used to transform $T_0(i, j)$ into the initial vector Y_0.
iii. Numerical integration is called using LSODA (for example).
iv. The function that defines the ODE system:
   a. Takes Y and converts it into the matrix T using reshape.
   b. All derivatives are computed in the matrix $\frac{\mathrm{d}T}{\mathrm{d}t}$ using finite differences in the original equation.
   c. The matrix $\frac{\mathrm{d}T}{\mathrm{d}t}$ is converted into a vector using reshape.
v. The matrix obtained at the integrator output is manipulated with reshape to transform its rows into matrices T at different integration times, to extract and visualize results.

Example 5.9 shows how to implement the method.

Example 5.9 Application of the method of lines to the development of the steady-state temperature profile in a solid with a square cross section

A steel bar with a rectangular cross section, 25 cm long and 20 cm high, is subject to the following boundary conditions:

$$T(0, y) = 100\ ^\circ\mathrm{C};\quad T(x, 0) = 50\ ^\circ\mathrm{C};\ (\mathrm{d}T/\mathrm{d}y)(y = 20) = 0;\quad 0 < x < 25$$
$$-k^*(\mathrm{d}T/\mathrm{d}x)(x = L) = h^*(T(x = L, y) - T_{amb}),\quad 0 < y < 20$$

- Data: $k = 0.1$ cal/(s cm °C), $h = 0.2$ cal/$\left(\text{s cm}^2\ ^\circ\text{C}\right)$, $\alpha = 0.038$ cm^2/s; $T_{\mathrm{amb}} = 15$ °C
- Initially, it is assumed that $T(t = 0) = T_{\mathrm{amb}} = 15$ °C
- Apply the method of lines and integrate until steady state is reached

In this example, Eq. (5.18) are complemented by the following expressions for the boundary conditions:

(1) BC for $i = 0$: $T_{0j} = 100$;
(2) BC for $i = N$: $(3+\mathrm{Bi})\cdot T_{N,j} - 4\cdot T_{N-1,j} + T_{N\_2j} = \mathrm{Bi}\cdot T_{\mathrm{amb}}$; $\mathrm{Bi} = 2\cdot\Delta x\cdot h/k$
(3) BC for $j = 0$: $T_{i_0} = 50$;
(4) BC for $j = M$: $T_{i,M-2} - 4\cdot T_{i,M-1} + 3\cdot T_{i,M} = 0$

Each time derivative of the boundary conditions gives rise to the ODE governing the temperature values at the boundaries. For example, the time derivative of the fourth boundary condition is equivalent to Eq. (5.18a), and so on.

Next, the numerical results and the corresponding code are presented.

```
import numpy as np
import matplotlib.pyplot as plt
from scipy.integrate import solve_ivp

def Steel(N, M, L, H, k, h, alpha, Tamb, tau):
    """
    Computes temperature profiles as a function of (x, y) in a rectangular
    steel bar of width L and height H (cgs units), with mixed boundary conditions.
    """

    # Spatial discretization
    dx = L / N
    dy = H / M          # dx = dy is assumed
    Nx = N + 1
    Ny = M + 1

    alpha_g = alpha / dx**2
    Bi_g = 2 * dx * h / k

    X = np.linspace(0, L, Nx)
    Y = np.linspace(0, H, Ny)

    # Initial condition
    T0 = Tamb * np.ones((Nx, Ny))
    T0[:, 0] = 50.0
    T0[0, 1:Ny] = 100.0
    T0[Nx-1, 1:Ny-1] = (
        4 * T0[Nx-2, 1:Ny-1] - T0[Nx-3, 1:Ny-1]
    ) / (3 + Bi_g)

    # Flatten initial condition (column-major, MATLAB-compatible)
    y0 = T0.T.reshape(Nx * Ny)

    # System of ODEs
    def system(t, y):
        T = y.reshape((Ny, Nx)).T
        Tp = np.zeros_like(T)

        for i in range(1, Nx-1):
            for j in range(1, Ny-1):
                Tp[i, j] = alpha_g * (
                    T[i+1, j] + T[i-1, j]
                    + T[i, j+1] + T[i, j-1]
                    - 4 * T[i, j]
                )

        return Tp.T.reshape(Nx * Ny)
```

```
    # Time integration (stiff solver, equivalent to ode15s)
    sol = solve_ivp(
        system,
        (tau[0], tau[-1]),
        y0,
        t_eval=tau,
        method="BDF"
    )

    # Temperature field at final time
    T = sol.y[:, -1].reshape((Ny, Nx)).T

    # Temperature at the center of the bar
    i1 = N // 2
    i2 = M // 2
    Tc = sol.y[(i1 * Ny + i2), :]

    # Contour plot at final time
    plt.figure()
    plt.contourf(X, Y, T.T, cmap="jet")
    plt.colorbar()
    plt.colormap = "jet"
    plt.xlabel("x (cm)")
    plt.ylabel("y (cm)")
    plt.title(f"Temperature (°C) at t = {tau[-1]} s")

    # Temperature at the center vs time
    plt.figure()
    plt.plot(sol.t, Tc, "-bo")
    plt.xlabel("time (s)")
    plt.ylabel("Temperature (°C)")
    plt.title("Temperature at the center of the bar")

    # Combined plot
    plt.figure(figsize=(10, 4))

    plt.subplot(1, 2, 1)
    plt.contourf(X, Y, T.T, cmap="jet")
    plt.colorbar()
    plt.xlabel("x (cm)")
    plt.ylabel("y (cm)")
    plt.title(f"Temperature (°C) at t = {tau[-1]} s")

    plt.subplot(1, 2, 2)
    plt.plot(sol.t, Tc, "-bo")
    plt.xlabel("time (s)")
    plt.ylabel("Temperature (°C)")
    plt.title("Temperature at the center of the bar")

    plt.tight_layout()
    plt.show()

    return T, Tc

# example 5.9 data
N=50
M=40
L=25
H=20
k=0.1
h=0.2
alpha=0.038
Tamb=15
tau=np.linspace(0, 4000, 100)
Steel(N, M, L, H, k, h, alpha, Tamb, tau);
```

5.6 Convection and Diffusion Problems in 2D

This Section addresses the problem of computing the distribution, in two spatial dimensions and one time dimension of variables such as temperature and concentration when a fluid velocity and pressure field is present. These equations arise from the analysis of dynamic transport problems in three spatial dimensions, which can often be well approximated by equations in two spatial dimensions and one temporal dimension. The classical example is the conservation equations of mass, energy, and momentum under the boundary-layer assumption, which makes it possible to reduce the system description by one spatial dimension. In addition, in many practical situations it is possible to approximate the variation along one of the spatial dimensions and treat it as essentially constant. Finally, when investigating a new problem in three spatial dimensions, it is often useful (and practical) to simplify the governing equations and first solve a two-dimensional problem to understand the main features of the numerical solution (Fig. 5.12).

One issue that is specific to this type of PDE has to do with the numerical treatment of convective terms using finite differences. The correct approach is to use upwind, first order differences for the spatial gradient multiplying the velocity vector. That is, using Eq. (4.5) instead of using a centered difference as in Eq. (4.6). For a detailed discussion of the justification for this approach, see Anderson et al. (2011).

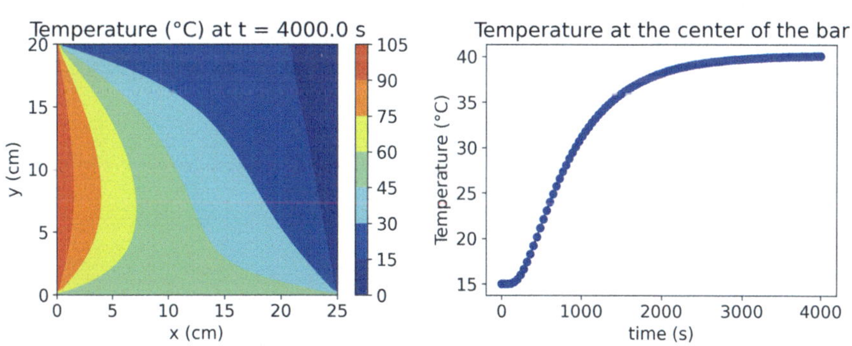

Fig. 5.12 Solution of Eq. (5.18) using the method of lines

5.6.1 Solution of the Flow Equations

The relevant equations in this case are the continuity equation and the Navier–Stokes equations in two spatial directions, which we choose as (x,y). The system of PDE to be solved is (Anderson et al., 2011):

$$\frac{\partial u}{\partial x} + \frac{\partial v}{\partial y} = 0 \tag{5.19}$$

$$\frac{\partial u}{\partial t} + u \cdot \frac{\partial u}{\partial x} + v \cdot \frac{\partial v}{\partial y} = -\frac{1}{\rho}\frac{\partial p}{\partial x} + \nu \cdot \left\{\frac{\partial^2 u}{\partial x^2} + \frac{\partial^2 u}{\partial y^2}\right\} \tag{5.20a}$$

$$\frac{\partial v}{\partial t} + u \cdot \frac{\partial v}{\partial x} + v \cdot \frac{\partial v}{\partial y} = -\frac{1}{\rho}\frac{\partial p}{\partial y} + \nu \cdot \left\{\frac{\partial^2 v}{\partial x^2} + \frac{\partial^2 v}{\partial y^2}\right\} \tag{5.20b}$$

To avoid computing pressure, this variable is eliminated from Eq. (5.20) by subtracting Eq. (5.20a) differentiated with respect to y minus Eq. (5.20b) differentiated with respect to x. In addition, the following functions are defined:

(a) The stream function ψ, defined by:

$$\frac{\partial \psi}{\partial x} = -v; \frac{\partial \psi}{\partial y} = u. \tag{5.21}$$

(b) The z-component of the vorticity vector, $\zeta = \nabla \times \mathbf{V}$, which corresponds to:

$$\zeta = \frac{\partial v}{\partial x} - \frac{\partial u}{\partial y}. \tag{5.22}$$

In this way, Eqs. (5.19) and (5.20) can be expressed, in terms of the two functions described above, by the following PDE:

$$\frac{\partial \zeta}{\partial t} + u \cdot \frac{\partial \zeta}{\partial x} + v \cdot \frac{\partial \zeta}{\partial y} = \nu \cdot \left\{\frac{\partial^2 \zeta}{\partial x^2} + \frac{\partial^2 \zeta}{\partial y^2}\right\} \tag{5.23}$$

$$\frac{\partial^2 \psi}{\partial x^2} + \frac{\partial^2 \psi}{\partial y^2} = -\zeta. \tag{5.24}$$

To solve these two PDE, the following sequential procedure is applied:

1. Specify the initial values of u, v, ζ, and ψ at $t = 0$.
2. Compute $\zeta(t + \Delta t)$ in the interior of the domain by integrating Eq. (5.23).
3. Solve the Poisson Eq. (5.24) for $\psi(t + \Delta t)$, using $\zeta(t + \Delta t)$.
4. Update the velocity components: $u = \partial\psi/\partial x$, $v = -\partial\psi/\partial y$.
5. Determine the values of ζ on the boundaries using the interior values of ζ and ψ.
6. Advance time and return to step (2).

If the pressure is to be evaluated, it can be derived from Eq. (5.20) in the form of a Poisson equation:

$$\frac{\partial^2 p}{\partial x^2} + \frac{\partial^2 p}{\partial y^2} = 2\rho\left\{\frac{\partial u}{\partial x}\frac{\partial v}{\partial y} - \frac{\partial u}{\partial y}\frac{\partial v}{\partial x}\right\}. \tag{5.25}$$

Returning to the sequential computational procedure, expressions are required for the boundary conditions satisfied by the functions ζ and ψ to continue advancing the numerical solution. First, if the flow boundaries are stationary walls, then the velocity vector is zero there and, therefore, ψ is a constant, i.e., independent of x and y, which can be assumed equal to zero. In this case, this function is kept always fixed at the flow boundary. On the other hand, in (x, y) coordinates, if a vertical wall is located at a fixed x position, then on that wall the horizontal velocity component is zero, i.e.,

$$u(x = x_1, y) = 0 = \frac{\partial\psi}{\partial y}; 0 < y < y_{\max} \to \frac{\partial^2\psi}{\partial y^2} = 0 \to \zeta = -\frac{\partial^2\psi}{\partial x^2}; 0 < y < y_{\max}. \tag{5.26}$$

Example 5.10 shows how to solve the square-cavity flow problem, including the specification of the boundary conditions.

Example 5.10 Solution of the dynamics of a two-dimensional laminar flow in a square-cavity domain

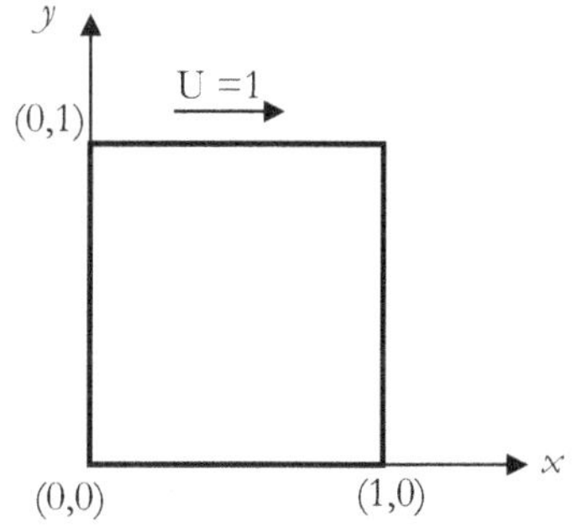

The figure shows a square cavity in which three boundaries are stationary, and the top boundary moves in the x-direction with a constant velocity $U = 1$. Initially, the fluid is at rest.

Determine the evolution of the velocity, vorticity, and stream function. The momentum diffusivity is $\nu = 0.1$. Thus, the Reynolds number is $Re = U \cdot 1/\nu = 10$, corresponding to laminar flow.

Initial conditions in the interior of the domain:

- The velocity field is zero.
- Consequently, the stream function is zero, according to Eq. (5.21).
- Likewise, the initial vorticity is zero, according to Eq. (5.22).

Boundary conditions on the domain boundary: the velocity components (u,v) are zero, except at the top boundary (y=1), where $u_{i,M} = U, i = 0, 1, \ldots, N$.

Consider the bottom boundary (y=0) and a Taylor expansion (in the y-direction) for the stream function ψ at a grid point i located just inside the domain:

$$\psi_{i,1} = \psi_{i,0} + \Delta y \cdot \left(\frac{\partial \psi}{\partial y}\right)_{y=0} + \frac{1}{2}(\Delta y)^2 \cdot \left(\frac{\partial^2 \psi}{\partial y^2}\right)_{y=0} + \mathcal{O}(\Delta y^3)$$

$$\psi_{i,1} = \psi_{i,0} + \Delta y \cdot u_{i,0} + \frac{1}{2}(\Delta y)^2 \cdot (-\zeta_{i,0}) + \mathcal{O}(\Delta y^3).$$

From this, the following relation is obtained:

$$\zeta_{i,0} = 2 \cdot \frac{(\psi_{i,0} - \psi_{i,1})}{(\Delta y)^2} + \mathcal{O}(\Delta y); i = 0, 1, \ldots, N.$$

Analogously, the following boundary conditions are obtained:

$$\zeta_{i,M} = 2 \cdot \frac{(\psi_{i,M} - \psi_{i,M-1} - U \cdot \Delta y)}{(\Delta y)^2} + \mathcal{O}(\Delta y); i = 0, 1, \ldots, N$$

$$\zeta_{0,j} = 2 \cdot \frac{(\psi_{0,j} - \psi_{1,j})}{(\Delta x)^2} + \mathcal{O}(\Delta x); j = 1, 2, \ldots, M - 1$$

$$\zeta_{N,j} = 2 \cdot \frac{(\psi_{N,j} - \psi_{N-1,j})}{(\Delta x)^2} + \mathcal{O}(\Delta x); j = 1, 2, \ldots, M - 1.$$

Thus, it is the motion of the top boundary of the cavity that generates vorticity, which in turn modifies the stream function, the velocity field, and so forth.

Figure 5.13 shows the result of the time integration of Eq. (5.23), followed by the solution of the Poisson Eq. (5.24) for ψ, etc., starting from the initial and boundary conditions specified above.

Next, the Python script for solving the transient problem up to $t = 1$ is presented. The reader is encouraged to run the code and determine the time required for the flow to reach the steady state, as in Example 5.9 above. Note that increasing the value of N leads to increasingly stiff ODE systems to be integrated.

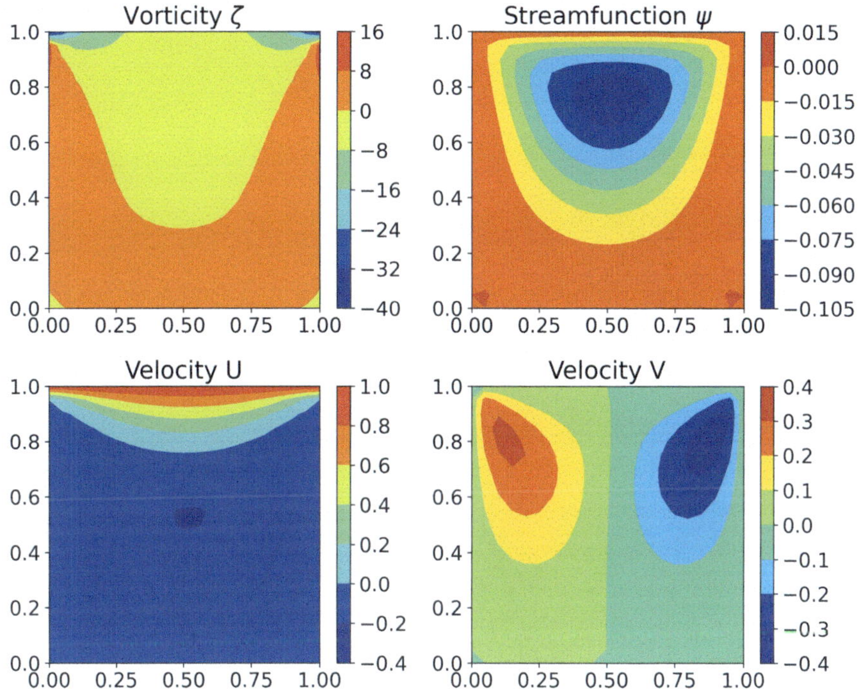

Fig. 5.13 Solution of Eq. (5.23) and Eq. (5.24) for laminar flow in a square cavity

```
import numpy as np
import matplotlib.pyplot as plt
from scipy.integrate import solve_ivp
from scipy.sparse import diags
from scipy.sparse.linalg import spsolve


def Flow_2D_CD(N, nu, Uw, tau):
    """
    Solves 2D incompressible flow in a square cavity using the
    vorticity-streamfunction formulation (central differences).
    """

    # Grid parameters
    dx = 1.0 / N                # dx = dy
    Nx = N + 1
    dx2 = dx**2
    nudx2 = nu / dx2

    X = np.linspace(0, 1, Nx)
    Y = np.linspace(0, 1, Nx)

    # SOR parameters for Poisson equation
    sigma = np.cos(np.pi / Nx)
    omega = 2.0 / (1.0 + np.sqrt(1.0 - sigma**2))

    # Tridiagonal matrix for Poisson solver
    e0 = np.ones(Nx)
    e1 = np.ones(Nx - 1)

    A = diags(
    diagonals=[-e1, 4*e0, -e1],
    offsets=[-1, 0, 1],
    shape=(Nx, Nx),
    format="csc"
).toarray()

# Boundary rows
    A[0, :] = 0
    A[0, 0] = 1
    A[-1, :] = 0
    A[-1, -1] = 1

    # Initialize fields
    U = np.zeros((Nx, Nx))
    V = np.zeros((Nx, Nx))
    Psi = np.zeros((Nx, Nx))
    Z = np.zeros((Nx, Nx))

    # Top wall boundary condition
    U[:, -1] = Uw
    Z[:, -1] = -2 * Uw / dx

    # ODE system for vorticity
    def system(t, y):
        Zloc = y.reshape((Nx, Nx)).T
        Zp = np.zeros_like(Zloc)

        for i in range(1, Nx-1):
            for j in range(1, Nx-1):
                diffusion = nudx2 * (
                    Zloc[i-1, j] + Zloc[i+1, j]
                    + Zloc[i, j-1] + Zloc[i, j+1]
                    - 4 * Zloc[i, j]
                )
                convection = (
                    U[i, j] * (Zloc[i, j] - Zloc[i-1, j]) / dx +
                    V[i, j] * (Zloc[i, j] - Zloc[i, j-1]) / dx
                )
                Zp[i, j] = diffusion - convection

        return Zp.T.reshape(Nx * Nx)

    # Poisson solver for streamfunction
    def poisson(Z):
        P = Psi.copy()
        err = 1.0
        it = 0
```

```
        while err > 1e-3 and it < 100:
            Pold = P.copy()

            for j in range(1, Nx-1):
                b = np.zeros(Nx)
                for i in range(1, Nx-1):
                    b[i] = dx2 * Z[i, j] + P[i, j-1] + Pold[i, j+1]

                v = np.linalg.solve(A, b)
                P[:, j] = Pold[:, j] + omega * (v - Pold[:, j])

            err = np.linalg.norm(P - Pold) / np.linalg.norm(P)
            it += 1

        return P

    # Time loop
    for k in range(1, len(tau)):
        print(f"Time step {k}")

        Z0 = Z.T.reshape(Nx * Nx)

        sol = solve_ivp(
            system,
            (tau[k-1], tau[k]),
            Z0,
            method="LSODA",
            rtol=1e-3,
            atol=1e-4
        )

        Z = sol.y[:, -1].reshape((Nx, Nx)).T

        Psi[:] = poisson(Z)

        # Update velocities
        for i in range(1, Nx-1):
            for j in range(1, Nx-1):
                U[i, j] = 0.5 * (Psi[i, j+1] - Psi[i, j-1]) / dx
                V[i, j] = -0.5 * (Psi[i+1, j] - Psi[i-1, j]) / dx

        # Boundary conditions for vorticity
        for i in range(Nx):
            Z[i, 0] = 2 * (Psi[i, 0] - Psi[i, 1]) / dx2
            Z[i, -1] = 2 * (Psi[i, -1] - Psi[i, -2] - Uw * dx) / dx2

        for j in range(1, Nx-1):
            Z[0, j] = 2 * (Psi[0, j] - Psi[1, j]) / dx2
            Z[-1, j] = 2 * (Psi[-1, j] - Psi[-2, j]) / dx2

    # Visualization
    plt.figure(figsize=(10, 8))

    plt.subplot(2, 2, 1)
    plt.contourf(X, Y, Z.T, cmap="jet")
    plt.colorbar()
    plt.title(r"Vorticity $\zeta$")

    plt.subplot(2, 2, 2)
    plt.contourf(X, Y, Psi.T, cmap="jet")
    plt.colorbar()
    plt.title("Streamfunction $\psi$")

    plt.subplot(2, 2, 3)
    plt.contourf(X, Y, U.T, cmap="jet")
    plt.colorbar()
    plt.title("Velocity U")

    plt.subplot(2, 2, 4)
    plt.contourf(X, Y, V.T, cmap="jet")
    plt.colorbar()
```

```
    plt.title("Velocity V")

    plt.tight_layout()
    plt.savefig("Figures_Flujo2D.png")
    plt.show()

    return U, V, Psi, Z

# generating Figure 5.13
N=20
nu=0.1
Uw=1
tau=np.linspace(0, 1, 100)
Flow_2D_CD(N, nu, Uw, tau);
```

5.6.2 Solution of the Momentum and Energy Balance Equations

Let us now consider the case of the simultaneous solution of Eqs. (5.23) and (5.24) together with the energy conservation Eq. (5.27), where α is the thermal diffusivity of the fluid:

$$\frac{\partial T}{\partial t} + u \cdot \frac{\partial T}{\partial x} + v \cdot \frac{\partial T}{\partial y} = \alpha \cdot \left\{ \frac{\partial^2 T}{\partial x^2} + \frac{\partial^2 T}{\partial y^2} \right\}. \tag{5.27}$$

The solution scheme is the same as previously proposed, except Eqs. (5.23) and (5.27) are solved simultaneously, given their similar structure. However, as changes appear in the spatial distribution of the fluid temperature, this induces changes in its density; consequently, the effect of the spatial variability of the fluid density arises, which generates natural convection. This is addressed using the Boussinesq approximation. Example 5.11 shows how to proceed in this case.

Example 5.11 Solution of the dynamics of two-dimensional laminar flow with natural convection in a rectangular-cavity domain

Consider a rectangular cavity as shown in the figure.

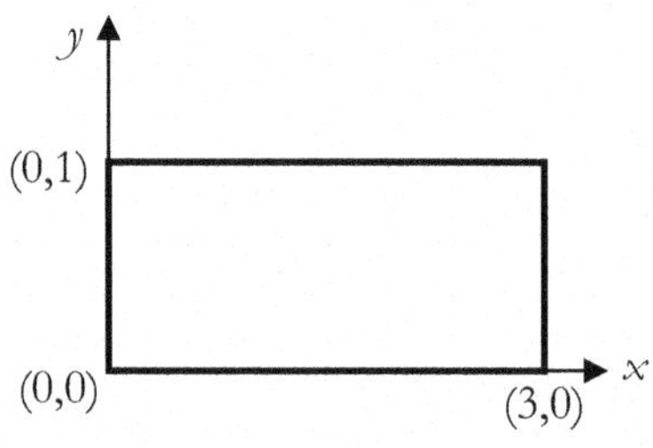

- Assume that $H = 1, L = 3, Pr = 7, Gr = 100$, and $\nu = 0.1$.

- The bottom boundary is set to $T = T_{\mathrm{H}}$ and the top boundary is cold at $T = T_{\mathrm{C}}$. The other boundaries are adiabatic.
- The top boundary is set in horizontal motion with velocity $u = U = 1$.
- The boundary conditions for $\phi = \frac{T-T_{\mathrm{C}}}{T_{\mathrm{H}}-T_{\mathrm{C}}}$ are:
- $x = 0 : \partial\phi/\partial x = 0; x = 3 : \partial\phi/\partial x = 0;$
- $y = 0 : \phi = 1; y = 1 : \phi = 0$
- Determine the steady-state solution for the velocity (u,v) and the temperature ϕ.

Due to the imposed vertical temperature gradient in the system, the vertical momentum equation, Eq. (5.20b), now becomes:

$$\frac{\partial v}{\partial t} + u \cdot \frac{\partial v}{\partial x} + v \cdot \frac{\partial v}{\partial y} = -\frac{1}{\rho}\frac{\partial p}{\partial y} + \nu \cdot \left\{\frac{\partial^2 v}{\partial x^2} + \frac{\partial^2 v}{\partial y^2}\right\} + g\beta\Delta T$$

where g is the acceleration due to gravity, β is the fluid volumetric expansion coefficient (units of °C^{-1}), and ΔT is the temperature difference of the fluid with respect to a reference state (T_{C} in this case). Using Eqs. (5.20a) and (5.20b) to eliminate the pressure, the following equation is obtained for the evolution of vorticity:

$$\frac{\partial \zeta}{\partial \tau} + u \cdot \frac{\partial \zeta}{\partial x} + v \cdot \frac{\partial \zeta}{\partial y} = \left\{\frac{\partial^2 \zeta}{\partial x^2} + \frac{\partial^2 \zeta}{\partial y^2}\right\} + \mathrm{Gr}\frac{\partial \phi}{\partial x}$$

where the following change of variables has been introduced: $\tau = (\nu\, t)/H^2$.

Substituting the dimensionless temperature ϕ into Eq. (5.27) yields:

$$\frac{\partial \phi}{\partial \tau} + u \cdot \frac{\partial \phi}{\partial x} + v \cdot \frac{\partial \phi}{\partial y} = \frac{1}{Pr} \cdot \left\{\frac{\partial^2 \phi}{\partial x^2} + \frac{\partial^2 \phi}{\partial y^2}\right\}$$

where Pr and Gr are the dimensionless Prandtl and Grashof numbers, respectively. The Prandtl number is defined as $\mathrm{Pr} = \nu/\alpha$, and it measures how effectively the fluid transports momentum relative to heat by diffusive processes in both cases. On the other hand, the Grashof number is defined as $\mathrm{Gr} = \left(g\beta\Delta T H^3\right)/\nu^2$; if Gr is large, this implies that buoyancy forces are dominant.

Figure 5.14 shows the results of the numerical integration of the previous two equations, together with Eq. (5.24) for the stream function ψ, up to $\tau = 3$.

Next, the Python script for solving Example 5.11 is presented. The reader is asked to verify whether a steady state has been reached at $\tau = 3$, or whether more time is required for this to occur. Also investigate the numerical solution as the Prandtl and Grashof numbers vary, and how the results change when the $\frac{L}{H}$ ratio is varied to values between 2 and 5, for example.

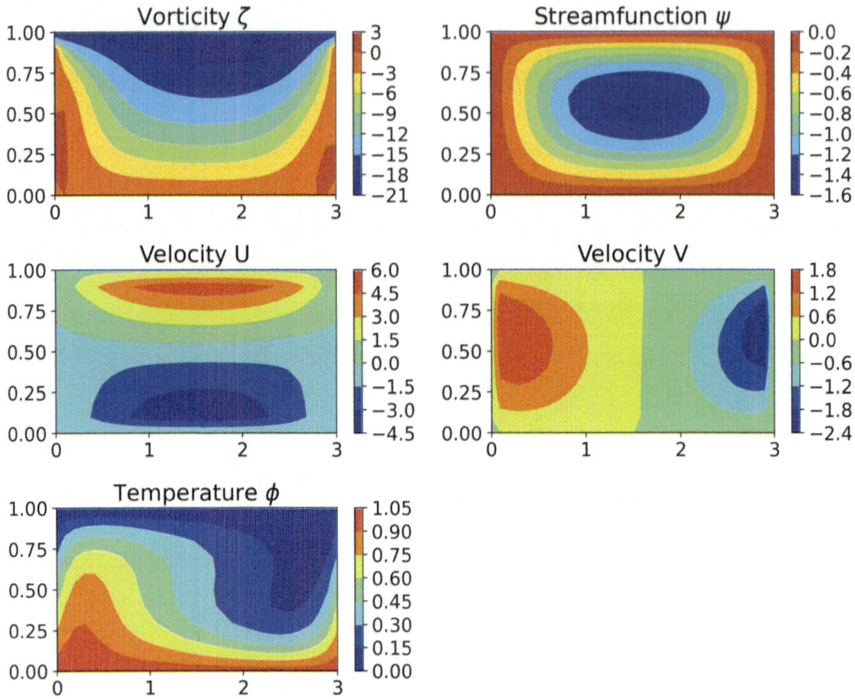

Fig. 5.14 Solution for Example 5.11

```
def Flow_2D_Temp2D_CD(N, M, Pr, Gr, Uw, tau):

    import numpy as np
    from scipy.integrate import solve_ivp

    # --- Grid and parameters ---
    dx = 3.0 / N
    dy = 1.0 / M
    dx2 = dx**2

    Nx = N + 1
    Ny = M + 1
    Nodes = Nx * Ny

    X = np.linspace(0, 3, Nx)
    Y = np.linspace(0, 1, Ny)

    Prdx2 = Pr * dx2
    G = Gr
```

```
    # --- Poisson solver parameters ---
    sigma = (np.cos(np.pi/N) + np.cos(np.pi/M)) / 2
    omega = 2 / (1 + np.sqrt(1 - sigma**2))

    n = Nx - 2
    a = -np.ones(n-1)
    b =  4*np.ones(n)
    c = -np.ones(n-1)

    # --- Fields ---
    U   = np.zeros((Nx, Ny))
    V   = np.zeros_like(U)
    Psi = np.zeros_like(U)
    Z   = np.zeros_like(U)
    phi = np.zeros_like(U)

    y = np.linspace(0, 1, Ny)

    # --- Boundary conditions ---
    for i in range(Nx):
        U[i, -1] = Uw
        Z[i, -1] = -2*Uw/dx
        phi[i, :] = 1 - y

    # ---------- helper functions (DEFINED FIRST) ----------

    def tridisolve(a, b, c, d):
        a = a.copy(); b = b.copy(); c = c.copy(); d = d.copy()
        n = len(d)

        for i in range(1, n):
            m = a[i-1] / b[i-1]
            b[i] -= m * c[i-1]
            d[i] -= m * d[i-1]

        x = np.zeros(n)
        x[-1] = d[-1] / b[-1]
        for i in range(n-2, -1, -1):
            x[i] = (d[i] - c[i] * x[i+1]) / b[i]

        return x

    def poisson(Z):
        nonlocal Psi

        err = 1.0
        while err > 1e-4:
            Pold = Psi.copy()

            for j in range(1, Ny-1):
                d = dx2*Z[1:-1, j] + Psi[1:-1, j-1] + Pold[1:-1, j+1]
                v = tridisolve(a, b, c, d)
                Psi[1:-1, j] = Pold[1:-1, j] + omega*(v - Pold[1:-1, j])

            err = np.linalg.norm(Psi - Pold) / np.linalg.norm(Psi)

        return Psi

    def sys(t, Y):
        Zloc = Y[:Nodes].reshape(Ny, Nx).T
        Tloc = Y[Nodes:].reshape(Ny, Nx).T

        Zp = np.zeros_like(Zloc)
        Tp = np.zeros_like(Tloc)

        for i in range(1, Nx-1):
            for j in range(1, Ny-1):
                Zp[i,j] = (
                    (Zloc[i-1,j] + Zloc[i+1,j] + Zloc[i,j-1] + Zloc[i,j+1] - 4*Zloc[i,j])
/ dx2
                    - U[i,j]*(Zloc[i,j]-Zloc[i-1,j])/dx
```

```
                - V[i,j]*(Zloc[i,j]-Zloc[i,j-1])/dy
                + G*(Tloc[i,j]-Tloc[i-1,j])/dx
            )

            Tp[i,j] = (
                (Tloc[i-1,j] + Tloc[i+1,j] + Tloc[i,j-1] + Tloc[i,j+1] - 4*Tloc[i,j])
/ Prdx2
                - U[i,j]*(Tloc[i,j]-Tloc[i-1,j])/dx
                - V[i,j]*(Tloc[i,j]-Tloc[i,j-1])/dy
            )

        return np.hstack([Zp.T.ravel(), Tp.T.ravel()])

    # ---------- time loop ----------

    for k in range(1, len(tau)):
        Y0 = np.hstack([Z.T.ravel(), phi.T.ravel()])

        sol = solve_ivp(
            sys, (tau[k-1], tau[k]), Y0,
            method="LSODA", rtol=1e-3, atol=1e-5
        )

        Z[:]   = sol.y[:Nodes, -1].reshape(Ny, Nx).T
        phi[:] = sol.y[Nodes:, -1].reshape(Ny, Nx).T

        Psi[:] = poisson(Z)
        print(k)
        for i in range(1, Nx-1):
            for j in range(1, Ny-1):
                U[i,j] = 0.5*(Psi[i,j+1] - Psi[i,j-1]) / dx
                V[i,j] =-0.5*(Psi[i+1,j] - Psi[i-1,j]) / dy

    return U, V, Psi, Z, phi

# Running example 5.11
import numpy as np
import matplotlib.pyplot as plt
N=30
M=10
Pr=7
Gr=100
Uw=1
tau=np.linspace(0, 3, 300)
U, V, Psi, Z, phi = Flow_2D_Temp2D_CD(N, M, Pr, Gr, Uw, tau)

# --- Grid and parameters ---
dx = 3.0 / N
dy = 1.0 / M
Nx = N + 1
Ny = M + 1
X = np.linspace(0, 3, Nx)
Y = np.linspace(0, 1, Ny)

# Visualization
plt.figure(figsize=(10, 8))

plt.subplot(3, 2, 1)
plt.contourf(X, Y, Z.T, cmap="jet")
plt.colorbar()
plt.title(r"Vorticity $\zeta$")

plt.subplot(3, 2, 2)
plt.contourf(X, Y, Psi.T, cmap="jet")
plt.colorbar()
plt.title("Streamfunction $\psi$")

plt.subplot(3, 2, 3)
plt.contourf(X, Y, U.T, cmap="jet")
plt.colorbar()
plt.title("Velocity U")
```

```
plt.subplot(3, 2, 4)
plt.contourf(X, Y, V.T, cmap="jet")
plt.colorbar()
plt.title("Velocity V")

plt.subplot(3, 2, 5)
plt.contourf(X, Y, phi.T, cmap="jet")
plt.colorbar()
plt.title("Temperature $\phi$")

plt.tight_layout()
plt.show()
```

5.7 Problems

(1) When a metallic cylinder conducts electricity, internal heat generation occurs, which raises the temperature of the metal; in this case, the effect of radiation must be considered. The following partial differential equation describes this phenomenon, using dimensionless variables:

$$\frac{\partial T}{\partial t} = \frac{\partial^2 T}{\partial r^2} + \frac{1}{r}\frac{\partial T}{\partial r} + G; \quad T(r, t = 0) = 1;$$

$$\text{(i)} \ \frac{\partial T}{\partial r}(r = 0, t \geq 0) = 0;$$

$$\text{(ii)} - \frac{\partial T}{\partial r}(r = 1, t \geq 0) = \text{Nu}\left\{[T(r = 1)]^4 - 1\right\}$$

Write a Python function that solves the previous equation and plots the temperature as a function of radius for different times, until steady state is reached. The plot must explicitly show the times you have chosen to present the solution (it is sufficient to plot 3 or 4 time points), the parameters $\{G, \text{Nu}\}$ used in the simulation, and that steady state has been reached (using different plotting symbols). Test your function with $G = 1$ and $\text{Nu} = 1$.

(2) Consider a spherical catalyst pellet (radius R) placed in a fluid containing a species at concentration C_0, which begins to diffuse into the interior of the pellet, where enzymatic degradation of the species occurs. The mass balance equation for the species concentration inside the pellet is given by:

$$\frac{\partial C}{\partial t} = D\left\{\frac{1}{r^2}\frac{\partial}{\partial r}\left(r^2\frac{\partial C}{\partial r}\right)\right\} - \frac{k \cdot C}{K + C}$$

where D is the diffusion coefficient, k is a kinetic constant, and K is a saturation constant. The previous equation is subject to the following conditions:

$$r = 0 : \left(\frac{\partial C}{\partial r}\right) = 0; \quad r = R : -D\left(\frac{\partial C}{\partial r}\right) = k_m(C - C_0)$$

$$t = 0 : \quad C = 0$$

Here, k_m is a mass-transfer coefficient between the pellet and the surrounding fluid.

(a) Use dimensionless variables to write the equation in a simpler form.
(b) Apply the method of lines to formulate the equations to be solved.
(c) Write Python routines that obtain the solution for the concentration C as a function of time and the radial coordinate. Include a graphical representation of the solution and how that plot should be interpreted.

(3) In a cylindrical reactor (length L, radius R), radial dispersion of a species occurs while the species reacts at the inner wall of the cylinder. Assuming a uniform velocity for the fluid entering the reactor, the mass balance equation for the compound reacting at the wall is given by

$$v_0 \frac{\partial C}{\partial z} = D\left\{\frac{1}{r}\frac{\partial}{\partial r}\left(r\frac{\partial C}{\partial r}\right)\right\}$$

Here, v_0 is the flow velocity and D is the radial dispersion coefficient inside the reactor. The previous equation is subject to the following conditions:

$$r = 0 : \left(\frac{\partial C}{\partial r}\right) = 0; \quad r = R : -D\left(\frac{\partial C}{\partial r}\right) = kC$$

$$z = 0 : \quad C = C_0$$

(a) Use dimensionless variables to write the equation in a simpler form.
(b) Write Python routines that obtain the solution for the concentration C as a function of the axial and radial coordinates. What would happen if a parabolic velocity profile were used, given by $V(r) = 2v_0\left[1 - (r/R)^2\right]$? Would anything change in your Python code? Explain.

(4) A cylindrical catalyst pellet (radius R) is placed in a fluid containing a species at concentration C_{ext}, which begins to diffuse into the interior of the pellet, where enzymatic degradation of the species occurs. The mass balance equation for the species concentration inside the pellet is given by

The previous equation is subject to the following conditions:

$$x = 0 : \left(\frac{\partial y}{\partial x}\right) = 0; \forall \tau \geq 0; \quad x = 1 : \left(\frac{\partial y}{\partial x}\right) = -Bi(y(x = 1) - 1); \forall \tau \geq 0$$

$$\tau = 0 : y = 0 \quad \forall x \in (0, 1)$$

(a) Program a routine that returns the profile $y(x, \tau)$ for any value of τ(passed as a function argument) and a value of N provided as an input argument to the routine.
(b) Use the previous routine to build a script that produces the profiles $y(x, \tau)$ for $\tau = 0.01, 0.05, 0.1$, and 0.5 on the same plot, using different symbols and appropriate titles.

(5) Consider the following parabolic equation:

$$\frac{\partial y}{dt} = \frac{\partial^2 y}{\partial x^2} \quad 0 < x < 1; t \geq 0$$

Subject to:

- Initial condition: $y(x, t = 0) = sen(\pi \cdot x)$
- Boundary conditions: $y(x = 0; t \geq 0) = 0; y(x = 1; t \geq 0) = 0$

Write a Python routine to compute (using the method of lines) the profile $y(x, t)$ for $0 \leq x \leq 1$ and $t = 0.005, 0.01, 0.02$, and 0.1, plotting all curves on the same figure and using distinctive symbols. Your routine must include the number N of nodes.

Also compute, at each of the times listed above, the absolute error of the method of lines at $x = 0.5$, by comparison with the analytical solution of the equation:

$$y_{EXACT}(x, t) = sen(\pi \cdot x) \cdot \exp\left(-\pi^2 \cdot t\right)$$

The routine must display the approximation errors.

(6) The following pair of partial differential equations represents the simultaneous diffusion of heat and mass within a spherical pellet of a porous solid when a first-order reaction occurs.

$$\frac{\partial c}{\partial t} = \frac{1}{r^2}\frac{\partial}{\partial r}\left(r^2 \frac{\partial c}{\partial r}\right) - \phi^2 \cdot f(c, T);$$

$$\frac{\partial c}{\partial r}(r = 0, t \geq 0) = 0; c(r = 1, t \geq 0) = 1; c(0 < r < 1, t = 0) = 0$$

$$\text{Le} \cdot \frac{\partial T}{\partial t} = \frac{1}{r^2}\frac{\partial}{\partial r}\left(r^2 \frac{\partial T}{\partial r}\right) + \beta \cdot \phi^2 \cdot f(c, T);$$

$$\frac{\partial T}{\partial r}(r = 0, t \geq 0) = 0; \quad T(r = 1, t \geq 0) = 1; \quad T(0 < r < 1, t = 0) = 0$$

$$f(c, T) = c \cdot \exp[\gamma \cdot (1 - 1/T)]$$

where c and T are the dimensionless concentration and temperature, respectively. The parameters ϕ^2, γ, and β correspond, respectively, to a reaction-rate parameter, an activation-energy parameter, and a heat-of-reaction parameter, all dimensionless. The parameter Le is the Lewis number of the pellet.

(a) Write a Python function that takes ϕ^2, γ, β, and Le as inputs to compute the radial concentration and temperature profiles inside the pellet as functions of time, where time is provided as a vector of integration times among the function arguments.
(b) Integrate the equations up to a time t at which steady state has been reached. Use as an initial set of times: [0.05 0.1 0.25 0.5 0.75 1.0].
(c) Your program must also plot the results for both profiles $(c(r), T(r))$ at different times and display the values of the function input parameters on the corresponding plots.

Use the following values to test your function:

(a) $\phi^2 = 1.1, \beta = 0.15, \gamma = 30$; Le $= 10$;
(b) $\phi^2 = 1.1, \beta = 0.15, \gamma = 30$; Le $= 1$;

(7) The temperature of the Earth's soil is governed by the mechanism of heat conduction in solids, which in one spatial dimension is characterized by the partial differential equation:

$$\frac{\partial T}{\partial t} = \alpha \cdot \frac{\partial^2 T}{\partial x^2}$$

where α is the thermal diffusivity of the soil ($\alpha = 0.085$ m^2/día), x is the depth below the ground surface, and t is time in days. The boundary conditions for the previous equation are given by:

$$x = H : T = T_{SM}, t \geq 0;$$

$$x = 0 : T = T_{SM} - \Delta T_S \cdot \cos\left[\frac{2 \cdot \pi}{365} \cdot (t - t_0)\right]; t \geq 0$$

In the previous equation, T_{SM} is the annual average ground-surface temperature, ΔT_S is its annual amplitude, and t_0 depends on the geographic location of the area.

(a) Solve the diffusion equation with the previous boundary conditions using the method of lines, employing the following values: $H = 30$ m, $N = 30$ discrete segments, $T_{SM} = 17.5$ °C, $\Delta T_S = 10.5$ °C, and $t_0 = 180$ days. Plot the temperatures at depths $x = 0$, 10, 20, and 30 m as functions of time.

(b) Verify that your results from (a) are sufficiently accurate by computing the solution with $N = 60$ discrete segments. Plot the temperatures at depths $x = 0$, 10, 20, and 30 m as functions of time, together with the temperatures computed in (a), on the same figure for comparison.

(8) Consider the case of an absorption column operating under plug-flow conditions with axial dispersion. If mass transfer between the liquid and solid phases is high, then adsorption equilibrium can be assumed to always hold. In this case, suppose that the adsorption isotherm is given by $n = (\gamma\, c)/(1 + K\, c)$, where n is the solute concentration in the solid and c is the solute concentration in the liquid; γ and K are the isotherm parameters. In this case, the mass conservation equation yields:

$$\left[1 + \frac{(1-\varphi)}{\varphi} \cdot \frac{dn}{dc}\right] \cdot \frac{\partial c}{\partial t} + \frac{\partial c}{\partial z} = \frac{1}{\mathrm{Pe}} \cdot \frac{\partial^2 c}{\partial z^2}; \quad 0 \le z \le 1; \quad t \ge 0$$

subject to the following boundary and initial conditions:

$$c(z=0) - \frac{1}{Pe}\frac{\partial c}{\partial z}(z=0) = 1$$

$$\frac{\partial c}{\partial z}(z=1) = 0$$

$$c(z, t=0) = \begin{cases} 0 \text{ if } 0 \le z < 0.25 \\ 1 \text{ if } 0.25 \le z < 0.75 \\ 0 \text{ if } 0.75 \le z \le 1 \end{cases}$$

where Pe is the Péclet number for the system and φ is the volumetric fraction occupied by the liquid phase.

(a) Solve the equations using the method of lines with finite differences. Use the following parameter values: $K = 2$, $\gamma = 3$, $\phi = 0.45$, $Pe = 50$, and dimensionless times t from 0 until steady state is reached (to be determined by trial in Python).

(b) Plot the different curves $n(z, t)$ and $c(z, t)$ in two separate figures for several discrete times as functions of z, using appropriate titles and axis labels. Also plot the liquid-phase concentration at $z = 1$ as a function of time. Approximately how long does it take for the system to reach steady state?

(9) Find the solution of Laplace's equation, $\nabla^2 T = 0$, in the rectangle R defined by $R = \{(x, y) : 0 \le x \le 4, 0 \le y \le 4\}$, where T is the temperature at an arbitrary point (x,y) in R. The boundary conditions in this case are:

$$\mathrm{T}(\mathrm{x}, 5) = 180 \ (0 \leq x < 4)$$
$$\mathrm{T}(4, \mathrm{y}) = 30 \quad (0 \leq y < 4)$$

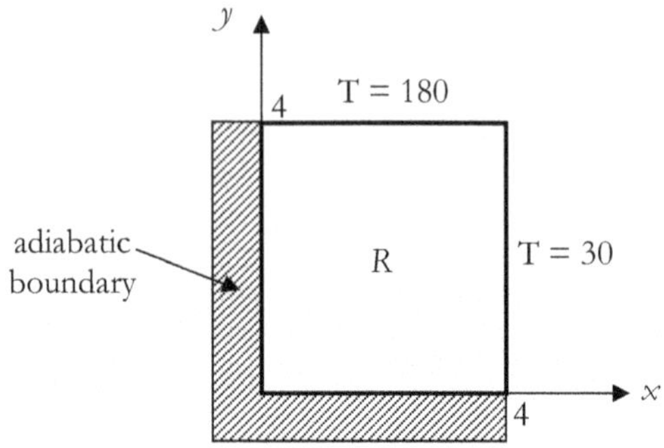

Using $\Delta x = \Delta y = 1/10$, solve the problem by comparing the iterative method without and with over-relaxation in terms of the number of iterations. Use relative tolerance of 10^{-3} for the norm of the solution T.

(10) Consider a metallic bar with a square cross section, 40 cm in height and 40 cm in width. The top wall is at 400 °C, the bottom wall at 60 °C, the left wall at 150 °C, and the right wall at 500 °C. The thermal diffusivity may be assumed constant, so the temperature distribution satisfies Laplace's equation, $\nabla^2 T = 0$.

(a) Compute the steady-state temperature distribution.
(b) Plot the temperature gradient for this steady-state distribution: $\nabla T = \left[\frac{\partial T}{\partial x} \frac{\partial T}{\partial y}\right]$
(c) How many discrete segments (N) are required so that the steady-state temperature at the center of the square profile no longer changes by more than 0.4 °C?

(11) In a cylindrical pellet, a chemical species diffuses; the pellet has been constructed with fibers, so diffusion is not the same in both directions, and the governing equation is

$$D_R\left\{\frac{1}{r}\frac{\partial}{\partial r}\left(r\frac{\partial C}{\partial r}\right)\right\} + D_Z\frac{\partial^2 C}{\partial z^2} = 0; \quad 0 < r < 1; \ 0 < z < 1$$

where D_R and D_Z are the radial and axial diffusion coefficients inside the pellet, respectively. The previous equation is subject to the boundary conditions

$$r = 0 : \left(\frac{\partial C}{\partial r}\right) = 0; \ r = 1 : C = 1$$
$$z = 0 : C = 0; \quad z = 1 : C = 2$$

Construct a Python function that solves this problem using the finite-difference method, with the same number of segments in both directions. Your function must generate a plot of the computed profile $C(r, z)$, with appropriate legends and titles. Test your function with the following cases.

(i) $D_R = 1$ and $D_Z = 1$;
(ii) $D_R = 0, 1$ and $D_Z = 1$;
(iii) $D_R = 1$ and $D_Z = 0.1$.

(12) A long catalyst pellet with a square cross section (side length $= 2L$) is placed in a fluid containing a species at concentration C_{ext}, which begins to diffuse into the pellet, where a first-order degradation of the species occurs. The steady-state mass balance equation for the species concentration inside the catalyst is given by:

$$\nabla^2 c = \frac{\partial^2 c}{\partial x^2} + \frac{\partial^2 c}{\partial y^2} = \phi^2 c; \quad 0 < x, y < 1$$

where $c(x, y) = C/C_{ext}$, $x = x'/L$, $y = y'/L$, and $\phi^2 = \frac{kL^2}{D}$(with D the diffusivity and k a kinetic constant), all dimensionless quantities. The previous equation is subject to the boundary conditions:

$$x = 0 : \left(\frac{\partial c}{\partial x}\right)(0, y) = 0; \; y = 0 : \left(\frac{\partial c}{\partial y}\right)(x, 0) = 0;$$

$$x = 1 : c(1, y) = 1; \; y = 1 : c(x, 1) = 1$$

Use the finite difference method to obtain the concentration within the porous pellet for $\phi^2 = 1$, 3, and 10. In each case, plot the concentration and also compute the pellet effectiveness factor: $\eta = \int_0^1 \int_0^1 c(x, y)\, dx\, dy$. Determine how many discrete segments (N) are required such that η no longer changes in its third significant digit.

(13) Consider a solid cylinder with radius $R = 1$ and length $L = 3$. The temperature profile inside the cylinder is given by the solution of the equation:

$$\left\{\frac{1}{r}\frac{\partial}{\partial r}\left(r\frac{\partial T}{\partial r}\right)\right\} + \frac{\partial^2 T}{\partial z^2} = 0; \quad 0 < r < 1; \quad 0 < z < 3$$

with the boundary conditions:

$$T(r = 1) = 50; \left(\frac{\partial T}{\partial r}\right)(r = 0) = 0; \, T(z = 0) = 20, \, T(z = 3) = 20$$

You are asked to use the finite difference method to compute the steady-state temperature profile $T(r, z)$. You are also asked to plot the temperature at the "center" of the domain ($r = 0.5$, $z = 1.5$) as a function of the number of iterations of the method. In both plots, indicate the value of the parameter ω ultimately used in the solution.

Notes:

(a) Use $N = 3M$ so that $\Delta z = \Delta r$.
(b) Use the first index i of the matrix for the z-direction, and the index j for the r-direction.
(c) Assume that the initial temperature is 0°C, except at the cylinder boundaries.
(d) Test different values of ω until the number of iterations is reduced for a given error (e.g., 10^{-3}); after you have estimated ω, reduce the error to a smaller value (e.g., 10^{-8}) to obtain the final results.

(14) Consider a thermal refining process in which heat is applied to a cylindrical solid so that the higher temperature promotes the thermal diffusion of certain undesirable elements (impurities) toward colder regions of the solid; thus, the heated zone gradually becomes free of them. The cylinder has radius $R = 2.5$ cm and length $L = 30$ cm. Heating is provided by an electrical resistance that supplies a constant heat flux q_0 at the left end of the cylinder ($z = 0$). The other end of the cylinder ($z = 30$) is insulated (adiabatic wall). The lateral surface of the cylinder is exposed to the surroundings, where heat exchange is convective with a known convection coefficient h. The heat conduction equation to be solved is as follows:

$$\rho C_p \frac{\partial T}{\partial t} = k\left\{\frac{1}{r}\frac{\partial}{\partial r}\left(r\frac{\partial T}{\partial r}\right) + \frac{\partial^2 T}{\partial z^2}\right\};$$

With the boundary conditions:

$$\left(\frac{\partial T}{\partial r}\right)(r = 0) = 0;\ -k \cdot \left(\frac{\partial T}{\partial r}\right)(r = R) = h(T(r = R) - T_{amb})$$

$$-k \cdot \left(\frac{\partial T}{\partial z}\right)(z = 0) = q_0;\quad \left(\frac{\partial T}{\partial z}\right)(z = L) = 0$$

and the initial condition that the temperature in the cylinder is uniform and equal to T_{amb}.

(a) Solve the heat conduction equation using the method of lines with the following values (cgs units): $\rho = 3$, $C_P = 0.3$, $h = 0.05$, $k = 1$, and $q_0 = 100$. Take $T_{\text{amb}} = 20$ °C.

(b) Plot the temperatures $T(r, z)$ at different times until an (approximately) steady state is reached. Also plot the temperature at the "center" of the cylinder ($r = 1.25$, $z = 15$) as a function of time. Plot the temperature profile along the cylinder axis of symmetry, $T(0, z)$, at the steady state obtained.

(15) Consider the development of the laminar velocity profile in a square duct of side length 2. In dimensionless form, the momentum equation is given by:

$$\frac{\partial v_z}{\partial \tau} = 1 + \frac{\partial^2 v_z}{\partial x^2} + \frac{\partial^2 v_z}{\partial y^2}; \quad -1 \le x \le 1; \ -1 \le y \le 1$$

subject to the boundary conditions: $v_z(x = \pm 1, y, t) = v_z(x, y = \pm 1, t) = 0$ and an initial condition of rest ($v_z(x, y, 0) = 0$).

(a) Apply the ADI method to integrate the previous equation until steady state is reached, and determine the steady-state distribution of the velocity $v_z(x, y)$. Include a plot of the velocity at the duct center to verify that steady state has been achieved.

(b) Compare the steady-state solution computed in (a) with the following approximate solution (Villadsen and Stewart, 1967):

$$\frac{v_z^{\mathrm{app}}(x, y)}{\left(1 - x^2\right)\left(1 - y^2\right)} = 0.31625 - 0.013125\left(2 - 5x^2 - 5y^2\right) + 0.004922\left(1 - 5x^2\right)\left(1 - 5y^2\right)$$

using a contour plot of the difference between the two velocities. What is the difference between the mean velocity of the two velocity distributions?

(16) When an elliptic equation contains nonlinear terms, it is simpler to solve it using the method of lines by adding a time derivative to the original equation and integrating until steady-state conditions are reached. Consider the equation Frank–Kamenetskii (1969):

$$\frac{\partial^2 \phi}{\partial x^2} + \frac{\partial^2 \phi}{\partial y^2} + \delta \cdot \exp(\phi) = 0 \quad ; \quad 0 < x < 1; 0 < y < 1;$$

Then, to find the solution to this equation, we use the method of lines, integrating until steady state is attained.

$$\frac{\partial \phi}{\partial \tau} = \frac{\partial^2 \phi}{\partial x^2} + \frac{\partial^2 \phi}{\partial y^2} + \delta \cdot \exp(\phi)$$

It is known that, for this equation, no solutions exist when $\delta > \delta_C \approx 1.7$. Plot the steady-state value $\phi(0,0)$ as a function of the parameter δ for $\delta = 0.1, 0.3, \ldots, 1.7$. In all cases, assume that initially $\phi = 0$. Also plot the solution $\phi(x, y)$ for the case $\delta = 1$.

(17) In 1952, Alan Turing published his theory of morphogenesis, in which he demonstrated the spontaneous generation of spatial patterns in systems with diffusion and chemical reaction. Consider the following reaction–diffusion model, described by the equations below (Binous et al., 2016):

$$\frac{\partial C_A}{\partial t} = D_A \cdot \left(\frac{\partial^2 C_A}{\partial x^2} + \frac{\partial^2 C_A}{\partial y^2}\right) + \frac{(C_A)^2}{C_B} - C_A; \quad 0 \leq x \leq 1; \ 0 \leq y \leq 1$$

$$\frac{\partial C_B}{\partial t} = D_B \cdot \left(\frac{\partial^2 C_B}{\partial x^2} + \frac{\partial^2 C_B}{\partial y^2}\right) + (C_A)^2 - C_B; \quad 0 \leq x \leq 1; \ 0 \leq y \leq 1$$

subject to no-flux (impermeability) boundary conditions:

$$y = 0 : \frac{\partial C}{\partial x} = 0; \ y = 1 : \frac{\partial C}{\partial x} = 0; \ x = 0 : \frac{\partial C}{\partial y} = 0; \ x = 1 : \frac{\partial C}{\partial y} = 0$$

and to initial conditions given by:

$$C_A(x, y, t = 0) = \text{rand}(\cdot); \ C_B(x, y, t = 0) = 0.1 + \text{rand}(\cdot);$$

where rand$(\cdot)$ denotes a random number uniformly distributed between 0 and 1, which is computed in Python using rng.random(). Consider the following values: $D_A = 0.001\text{cm}^2/\text{s}$; $D_B = 0.01\text{cm}^2/\text{s}$.

(a) Apply the method of lines to integrate the preceding differential equations up to $t = 100$.
(b) For $t = 0$ and $t = 100$, plot the concentrations C_A and C_B, as well as the time evolution of these concentrations at the center (0.5, 0.5). Indicate whether steady state has been reached at that point.
(c) Is the formation of any spatially ordered structure in C_A or C_B observed? Or is there a tendency toward uniform concentration throughout the spatial domain?

(18) Repeat the analysis of Example 5.10 but now change the boundary conditions: the top boundary moves with velocity $U = 1/2$, while the bottom boundary moves with velocity $U = -1/2$.

(a) How long does the system take to reach steady state?
(b) Compare your results using the LSODA integrator or, alternatively, using the explicit Euler method to solve the system of differential equations. What time-integration step is required to obtain results within 1% error relative to ode15s?

(19) Repeat the analysis of Example 5.10 but now assume that a solute is diffusing into the cavity, which is always maintained isothermal. The solute concentration C satisfies the equation:

$$\frac{\partial C}{\partial t} + u \cdot \frac{\partial C}{\partial x} + v \cdot \frac{\partial C}{\partial y} = D \cdot \left\{ \frac{\partial^2 C}{\partial x^2} + \frac{\partial^2 C}{\partial y^2} \right\}$$

with no-flux (impermeability) conditions on all cavity walls:

$$y = 0 : \frac{\partial C}{\partial x} = 0; y = 1 : \frac{\partial C}{\partial x} = 0; x = 0 : \frac{\partial C}{\partial y} = 0; x = 1 : \frac{\partial C}{\partial y} = 0$$

and with the initial condition:

$$t = 0 : C(x, y) = y; 0 \leq x \leq 1; 0 \leq y \leq 1.$$

Obtain solutions for the velocity distribution, stream function, vorticity, and concentration C, and plot them for different final times $\tau = 1, 2,$ and 3.

(a) Estimate the time required for the cavity center to reach steady state in velocities (u, v) and concentration (for a fixed value of N).
(b) Estimate how many discrete segments (N) would be required so that the concentrations and velocities at the cavity center do not change by more than 1% at the final time you estimated in (a).

Consider the following parameter values: $\nu = 0.1$, $D = 0.02$. In this case, the Schmidt number appears in the concentration transport equation, where $\text{Sc} = \nu/D$.

(20) Repeat the analysis of Example 5.11, but now include solute diffusion as described in the previous problem. There are now three convection–diffusion equations: vorticity, temperature, and concentration. Estimate how long it takes the system to reach steady state. To verify that your code is correct, compare your results with those obtained using the explicit Euler method applied to these three equations with a small step.

References

Anderson, D. A., Tannehill, J. C., & Pletcher, R. H. (2011). *Computational fluid mechanics and heat transfer* (3rd ed.). CRC Press.

Aris, R. (1999). *Mathematical modeling*. Academic Press.

Binous, H., Kaddeche, S., & Bellagi, A. (2016). Solving two-dimensional chemical engineering problems using the Chebyshev orthogonal collocation technique. *Computer Application in Engineering Education, 24*, 144–155. https://doi.org/10.1002/cae.21680

Crank, J., & Nicolson, P. (1947). A practical method for numerical evaluation of solutions of partial differential equations of the heat conduction type. *Proceedings of the Cambridge Philological Society, 43*, 50–67.

Finlayson, B. (2002). Numerical Methods for Chemical Engineers. http://faculty.washington.edu/finlayso/ebook/. Accessed 10 January 2026.

Frank-Kamenetzkii, D. A. (1969). *Diffusion and heat transfer in chemical kinetics*. Plenum Press.

Kee, R. J., & Petzold, L. (1986). A differential/algebraic equation formulation of the method-of-lines solution to systems of partial differential equations, Sandia National Lab report SAND86-8893. https://cse.cs.ucsb.edu/sites/default/files/publications/SAND868893.pdf. Accessed 10 January 2026.

Moler, C. (2008). *Numerical computing with MATLAB*, http://www.mathworks.com/moler/. Accessed 10 January 2026.

Rice, R. G., & Do, D. D. (2012). *Applied Mathematics and Modeling for Chemical Engineers*, 2nd Edition. Wiley.

Sewell, G. (1988). *The numerical solution of ordinary and partial differential equations*. Academic Press.

Turing, A. (1952). The chemical basis of morphogenesis. *Philos Transaction of Royal Society London, B237*, 37–72. https://doi.org/10.1098/rstb.1952.0012

Villadsen, J. V., & Stewart, W. E. (1967). Solution of boundary-value problems by orthogonal collocation. *Chemical Engineering Science, 22*, 1483–1501.

Weinberger, S. (1995). *A first course in partial differential equations: With complex variables and transform methods*. Dover.

Part II
Applied Problems in Chemical Process Engineering

Chapter 6
Case Studies

6.1 Problem 1 Multiple Reactions in a Batch Reactor

In a batch reactor, multiple liquid-phase reactions take place. The reaction scheme can be characterized as follows:

$$A \leftrightarrow B \rightarrow C$$

where k_{1f} ($A \rightarrow B$), k_{1r} ($B \rightarrow A$) and k_2 ($B \rightarrow C$) denote the kinetic rate constants. Thus, component A reacts reversibly to form the desired compound B. However, compound B can further react to form the undesired compound C.

Assuming first-order kinetics and constant reactor volume, determine:

(a) For $k_{1f} = 2\ \text{h}^{-1}$, $k_{1r} = 1\ \text{h}^{-1}$, $k_2 = 1.25\ \text{h}^{-1}$, $C_{A0} = 1$ mol/L, $C_{B0} = C_{C0} = 0$. Determine the time evolution of the concentrations of species A, B, and C over a 5-h period. Identify the time at which the intermediate species B attains its maximum concentration and determine the corresponding concentration value.

(b) In real processes, some uncertainty typically exists in the kinetic rate constants. Assume therefore that the new "actual" value of k_2 is $1.5\ \text{h}^{-1}$. Quantitatively compare the new results with those obtained in part (a).

Solution

This idealized situation requires three mass balances (later we will show that two differential equations are sufficient). Since the volume in which the reactions take place is constant, we can express the mass balance as:

$$\frac{\mathrm{d}C_A}{\mathrm{d}t} = k_{1r} \cdot C_B - k_{1f} \cdot C_A$$
$$\frac{\mathrm{d}C_B}{\mathrm{d}t} = k_{1f} \cdot C_A - k_2 \cdot C_B$$

H. Jorquera and C. A. Gelmi, *Numerical Methods in Chemical Process Engineering Using Python*, https://doi.org/10.1007/978-3-032-22958-8_6

Table P1.1 Comparison of results, parts (a) and (b)

	$k_2 = 1.25$	$k_2 = 1.5$	Δ%
Maximum value of *B* (mol/L)	0.378	0.354	– 6.3
Time at which the maximum is reached (h)	0.566	0.526	– 7.1

$$\frac{dC_C}{dt} = k_2 \cdot C_B \tag{P1.1}$$

The following profiles are obtained by integrating the system of differential equations:

For part (b), the graphical solution is very similar, and therefore the figure is omitted. Instead, the numerical results for both cases are compared in Table P1.1.

As can be observed, with a 20% error in the reaction rate constant k_2, the maximum value of B decreases by 6.3%. Furthermore, the time at which this maximum is reached varies by 7.1% with respect to the base case (part (a)). That is, the variable most affected by the uncertainty in parameter k_2 is the time at which the maximum occurs.

The Python code developed in this case was:

```
import numpy as np
from scipy.integrate import solve_ivp
import matplotlib.pyplot as plt

# System of differential equations and constants
def probl(t, c):
    # Parameters
    k1f = 2.0    # 1/h
    k1r = 1.0    # 1/h
    k2 = 1.25    # 1/h

    # Model
    dc1dt = -k1f*c[0] + k1r*c[1]
    dc2dt = k1f*c[0] - (k1r + k2)*c[1]
    dc3dt = k2*c[1]
    return [dc1dt, dc2dt, dc3dt]

# Initial conditions
c0 = [1.0, 0.0, 0.0]

# Time interval and specific time points
t_span = (0, 5)
t = np.linspace(0, 5, 1000)  # The solution will be evaluated at these points

# Model integration
sol = solve_ivp(probl, t_span, c0, t_eval=t, method='RK45')
CA, CB, CC = sol.y
```

```
# Plotting
plt.figure(figsize=(8, 6))
plt.plot(t, CA, 'k', label='A')
plt.plot(t, CB, 'k', label='B', lw=3)
plt.plot(t, CC, 'k--', label='C')
plt.xlabel('Time (h)', fontsize=14)
plt.ylabel('Concentration (mol/L)', fontsize=14)
plt.xlim(0, 5)
plt.ylim(0, 1)
plt.legend()
plt.tick_params(axis='both', labelsize=14)
plt.show()

# Computes the maximum concentration of B and the time required to reach it
max_B = np.max(CB)
pos_B = np.argmax(CB)
tpo_max_B = t[pos_B]
print(f"Maximum concentration of B: {round(max_B, 3)}")
print(f"Time to reach the maximum concentration of B: {round(tpo_max_B, 3)} hours.")
```

To obtain the concentration profiles, the system of differential equations was integrated using the `solve_ivp`[1] function from the SciPy library. This integrator solves initial-value problems for systems of ordinary differential equations (ODE) and provides a flexible interface that supports various integration methods for both stiff and non-stiff systems (including Runge–Kutta methods of different orders, multi-step BDF schemes, and Adams/BDF). In addition, it allows the user to control key aspects such as accuracy and the integration step size, among other parameters.

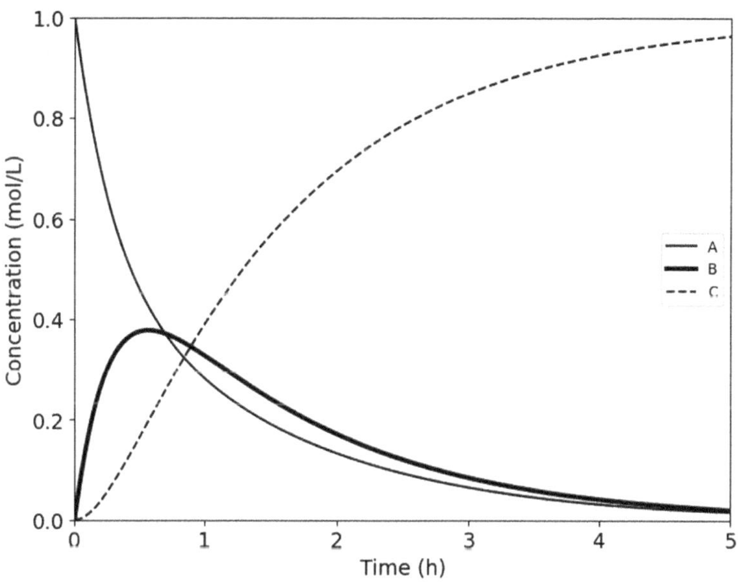

Fig. P1.1 Concentration as a function of time, part (a)

[1] For further details, consult the documentation: https://docs.scipy.org/doc/scipy/reference/generated/scipy.integrate.solve_ivp.html.

Once the differential system was integrated, the maximum values of compound B were obtained using the `max` and `argmax` functions from the NumPy library, which return the maximum value and its position in the array or vector, respectively. A reasonably close result could have been obtained by directly reading the values from Fig. P1.1.

6.2 Problem 2 Optimal Residence Time for Series Reactions in a CSTR

The following elementary first-order liquid-phase reactions take place in a continuous stirred tank reactor (CSTR):

$$2A \xrightarrow{k_1} B \xrightarrow{k_2} C$$

The kinetic rate constants are $k_1 = 1.0\ 1/\text{mol min}$ and $k_2 = 0.1\ \text{min}^{-1}$. Assume that the inlet concentration of compound A is $C_{A0} = 1\ \text{mol/L}$ and $C_{B0} = C_{C0} = 0$.

(a) Compute the steady-state concentrations of compounds *A*, *B*, and *C* for a residence time (τ) of 0.5 min.
(b) Plot the concentration of compound *B* as a function of the residence time τ. Determine the maximum value reached by compound *B*.

Solution

The differential mass balances for the most general case are given by:

$$\begin{aligned}\frac{dC_A}{dt} &= \frac{C_{A0} - C_A}{\tau} - 2 \cdot k_1 \cdot C_A^2 \\ \frac{dC_B}{dt} &= \frac{C_{B0} - C_B}{\tau} + k_1 \cdot C_A^2 - k_2 \cdot C_B. \\ \frac{dC_C}{dt} &= \frac{C_{C0} - C_C}{\tau} + k_2 \cdot C_B\end{aligned} \tag{P2.1}$$

The residence time (τ) is defined as the ratio between the liquid-phase volume of the reactor (V) and the inlet volumetric flow rate (v_0). The terms C_{B0} and C_{C0} in the above equations are zero because these compounds are not present in the inlet stream.

To compute the steady-state concentrations, the time derivatives are set equal to zero. After doing so, one must solve a system of three nonlinear equations with three unknowns (C_A, C_B, and C_C).

The Python codes for both parts of the problem are:

```
import numpy as np
from scipy.optimize import fsolve
import matplotlib.pyplot as plt

# System of three nonlinear equations
def prob2(c, tau):
    # Parameters
    k1 = 1.0    # 1/min
    k2 = 0.1    # 1/min
    ca0 = 1.0   # mol/L

    # Equations
    eq1 = (ca0 - c[0])/tau - 2*k1*c[0]**2
    eq2 = -c[1]/tau + k1*c[0]**2 - k2*c[1]
    eq3 = -c[2]/tau + k2*c[1]
    return [eq1, eq2, eq3]

# Part (a)
print('- Part (a):')
c_a, _, _, mesg = fsolve(prob2, [1, 1, 1], args=(0.5,), xtol=1e-8, full_output=True)
print("Convergence message:", mesg)
print("Solution for part (a):", np.round(c_a, 3))

# Part (b)
print()
print('- Part (b):')
i = 0
c0_b = [1, 1, 1]
tau = np.arange(0.1, 20, 0.01)
c_b = np.zeros((len(tau), 3))

for taui in tau:
    c_b[i, :] = fsolve(prob2, c0_b, args=(taui,), xtol=1e-8)
    c0_b = c_b[i, :]
    i += 1

# Plot for part (b)
plt.figure(figsize=(8, 6))
plt.plot(tau, c_b[:, 1], 'k')
plt.xlabel('τ (min)', fontsize=14)
plt.xlim(0, 20)
plt.ylabel('Concentration (mol/L)', fontsize=14)
plt.ylim(0.05,)
plt.tick_params(axis='both', labelsize=14)
plt.show()

# Compute maximum value and corresponding residence time
max_B = np.max(c_b[:, 1])
pos_B = np.argmax(c_b[:, 1])
tau_max_B = tau[pos_B]

print("Maximum concentration of B:", round(max_B, 3))
print("Residence time at maximum:", round(tau_max_B, 2), "minutes.")
```

The `fsolve` function from the SciPy library is used to compute the roots of the 3 × 3 system of nonlinear equations. This implies that the objective is to determine the set of values for which the system becomes equal to zero. By enabling the `full_output` option in `fsolve`, one can access the termination message (`mesg`), which, if a solution is not found, provides detailed information regarding the reason for non-convergence.

In the second part of the problem, a vector of τ values is constructed, ranging from 0.1 to 20 with a step size of 0.01. For each value of τ, the system of equations is solved using the `fsolve` routine, and the resulting solution is stored in the matrix `c_b`. At each call to `fsolve`, the steady-state solution obtained in the previous iteration is used as the initial guess, `c0_b`. This continuation strategy improves the convergence of the numerical solver.

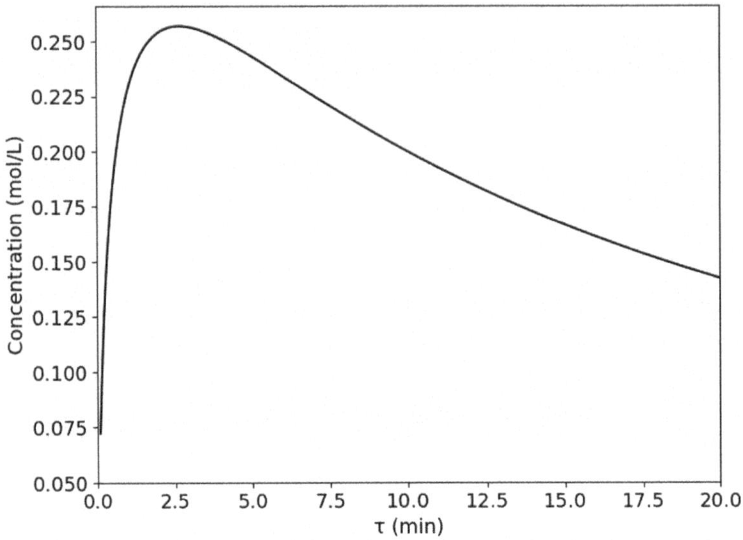

Fig. P2.1 Steady-state concentration of compound B versus residence time τ

Accordingly, for part (a), the program yields the following steady-state values:

$$\mathbf{C_A} = 0.618(\text{mol/L});\ \mathbf{C_B} = 0.182(\text{mol/L});\ \mathbf{C_C} = 0.009(\text{mol/L}).$$

For part (b), Fig. P2.1 summarizes all steady-state concentrations as a function of τ.

Once again, by applying NumPy's `max` and `argmax` functions to the concentration vector (c), the following results are obtained:

$$\begin{aligned} \mathbf{C_B max} &= 0.257\text{mol/L} \\ \tau(\mathbf{C_B max}) &= 2.68\text{min} \end{aligned}.$$

6.3 Problem 3 CSTRs in Series with Transport Delay

Consider the system shown in Fig. P3.1, in which a transport delay (D) (e.g., a very long pipeline) exists between reactor 1 and reactor 2.

Data: k_{ij} denote the kinetic constant for reaction i ($i = 1, 2$) occurring in reactor j ($j = 1, 2$). Reactors 1 and 2 have volumes $V_1 = 200$ L and $V_2 = 300$ L, respectively. At $t = 0$, the concentrations of species B and C in both reactors are zero, i.e., $C_B = C_C = 0$. In reactor 1, the initial concentration of species A is $C_{A0} = 8.5$ mol/L, and the feed concentration is $C_{Af} = 0.5$ mol/L. In reactor 2, the initial concentration of

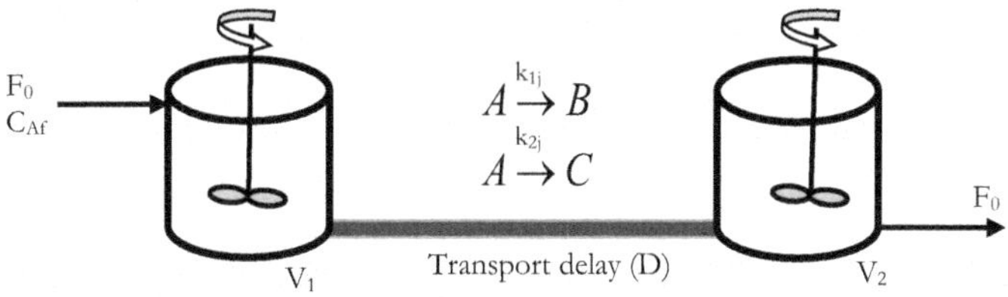

Fig. P3.1 Reactor configuration scheme

species A is $C_{A0} = 1$ mol/L. The volumetric flow rate is $F_0 = 300$ L/h. The kinetic constants are $k_{11} = k_{12} = 4\ \text{h}^{-1}$, $k_{21} = k_{22} = 3.2\ \text{h}^{-1}$.

Assumptions:

(i) No reactions occur in the pipeline connecting the reactors.
(ii) Both reactors are liquid-filled.
(iii) At t = 0, the pipeline contains inert compounds only. It contains inert compounds.

Based on this information:

(a) Determine a mathematical model capable of describing the dynamics of the entire system.
(b) Plot the concentrations of species A, B, and C as functions of time for both reactors, over a 30-min time horizon, considering a transport delay $D = 6$ min.

Solution

The presence of a transport delay between the two reactors introduces additional complexity. Without this delay, the numerical solution would be straightforward; however, the inlet stream to the second reactor must be treated explicitly.

The mass balance for the first reactor is:

$$\begin{aligned} \frac{dC_{A1}}{dt} &= \frac{F_0 \cdot (C_{A0} - C_{A1})}{V_1} - (k_{11} + k_{21}) \cdot C_{A1} \\ \frac{dC_{B1}}{dt} &= k_{11} \cdot C_{A1} - \frac{F_0 \cdot C_{B1}}{V_1} \\ \frac{dC_{C1}}{dt} &= k_{21} \cdot C_{A1} - \frac{F_0 \cdot C_{C1}}{V_1} \end{aligned} \quad . \tag{P3.1}$$

Because the volumetric flow rate is constant throughout the system, the reactor volumes are constant. Therefore, the right-hand side of Eq. (P3.2) can be divided by the volume of the corresponding reactor.

For the second reactor, the mass balance is somewhat more interesting. Here, two cases must be distinguished:

(I) **Case 1: $t < D$**

$$\begin{aligned}\frac{dC_{A2}}{dt} &= -\frac{F_0 \cdot C_{A2}}{V_2} - (k_{12} + k_{22}) \cdot C_{A2} \\ \frac{dC_{B2}}{dt} &= k_{12} \cdot C_{A2} - \frac{F_0 \cdot C_{B2}}{V_2} \\ \frac{dC_{C2}}{dt} &= k_{22} \cdot C_{A2} - \frac{F_0 \cdot C_{C2}}{V_2}\end{aligned} \tag{P3.2}$$

(ii) **Case 2: $t \geq D$**

$$\begin{aligned}\frac{dC_{A2}}{dt} &= \frac{F_0 \cdot \left(C_{A1}|_{(t-D)} - C_{A2}\right)}{V_2} - (k_{12} + k_{22}) \cdot C_{A2} \\ \frac{dC_{B2}}{dt} &= \frac{F_0 \cdot \left(C_{B1}|_{(t-D)} - C_{B2}\right)}{V_2} + k_{12} \cdot C_{A2} \\ \frac{dC_{C2}}{dt} &= \frac{F_0 \cdot \left(C_{C1}|_{(t-D)} - C_{C2}\right)}{V_2} + k_{22} \cdot C_{A2}\end{aligned} \tag{P3.3}$$

The term $C_{i1}|_{(t-D)}$ corresponds to the concentration of compound i in the first reactor evaluated at $(t - D)$. However, Eq. (P3.3) introduce an additional difficulty: they require the concentration evaluated at a time shifted relative to the current integration time, i.e., a term of the form $C_{i1}|_{(t-D)}$. To overcome this difficulty, several strategies may be employed. The simplest approach is to use the `interp` function from the NumPy library, which allows interpolation of values in a data vector (x, y). The idea is to obtain an interpolated value of the concentrations of compounds A, B, and C at the outlet of the first reactor as a function of time. This allows the integrated system to evaluate the outlet concentrations of the first reactor at any required time, in particular, at $t - D$.

Other alternatives include the use of integrators specifically designed for delay differential equations. In this problem, however, we will adopt the first strategy.

Below is the Python code:

```
import numpy as np
import matplotlib.pyplot as plt
from scipy.integrate import solve_ivp

def modelo1(t, c):
    # Parameters for Reactor 1
    V1 = 200; F0 = 300
    k11 = 4; k12 = 4
    k21 = 3.2; k22 = 3.2
    CAf = 0.5

    # Mass balance for compound A in Reactor 1
    dc1 = F0/V1*(CAf - c[0]) - (k11 + k21)*c[0]
    dc2 = -F0/V1*c[1] + k11*c[0]
    dc3 = -F0/V1*c[2] + k21*c[0]
    return [dc1, dc2, dc3]

def modelo2(t, c, C1lag):
    # Parameters for Reactor 2
    V2 = 300; F0 = 300
    k12 = 4; k22 = 3.2
    CAf = 0.5; D = 0.1  # 1/h

    # Conditions before and after the delay
    if t - D > 0:
        CA1lag = np.interp(t - D, C1lag[:, 0], C1lag[:, 1])
        CB1lag = np.interp(t - D, C1lag[:, 0], C1lag[:, 2])
        CC1lag = np.interp(t - D, C1lag[:, 0], C1lag[:, 3])
    else:
        CA1lag = 0
        CB1lag = 0
        CC1lag = 0

    # Mass balance for compound A in Reactor 2
    dc1 = F0/V2*(CA1lag - c[0]) - (k12 + k22)*c[0]
    dc2 = F0/V2*(CB1lag - c[1]) + k12*c[0]
    dc3 = F0/V2*(CC1lag - c[2]) + k22*c[0]
    return [dc1, dc2, dc3]

# Initial conditions
c1_inicial = [8.5, 0, 0]
c2_inicial = [1, 0, 0]

# Time range for integration (hours)
time_span = (0, 0.5)
t = np.linspace(0, 0.5, 100)

# Reactor 1 integration
C1 = solve_ivp(modelo1, time_span, c1_inicial, t_eval=t, method='LSODA')
RC1 = np.column_stack((C1.t, C1.y.T))

# Reactor 2 integration
C2 = solve_ivp(modelo2, time_span, c2_inicial, t_eval=t, args=(RC1,), method='LSODA')

# Extract concentration profiles
CA1, CB1, CC1 = C1.y
CA2, CB2, CC2 = C2.y

# Plotting
plt.figure(figsize=(8, 6))
```

```
plt.subplot(2, 1, 1)
plt.plot(t, CA1, 'k', lw=2, label=r'C$_{A}$')
plt.plot(t, CB1, 'k-.', lw=1, label=r'C$_{B}$')
plt.plot(t, CC1, 'k--', lw=1, label=r'C$_{C}$')
plt.title('Reactor 1')
plt.ylabel('Concentration (mol/L)', fontsize=14, labelpad=20)
plt.xlim(0, 0.5)
plt.tick_params(axis='both', labelsize=14)
plt.legend()

plt.subplot(2, 1, 2)
plt.plot(t, CA2, 'k', lw=2, label=r'C$_{A}$')
plt.plot(t, CB2, 'k-.', lw=1, label=r'C$_{B}$')
plt.plot(t, CC2, 'k--', lw=1, label=r'C$_{C}$')
plt.axvline(x=0.1, color='k', linewidth=3, alpha=0.5)
plt.text(0.11, 1, 'Arrival of A, B, and C\nto Reactor 2', fontsize=9, color='k',
weight='bold')
plt.title('Reactor 2')
plt.xlabel('Time (h)', fontsize=14)
plt.xlim(0, 0.5)
plt.ylabel('Concentration (mol/L)', fontsize=14, labelpad=6)
plt.tick_params(axis='both', labelsize=14)
plt.legend()

plt.tight_layout()
plt.show()
```

In this case, we used the `solve_ivp` integrator from the SciPy library. This routine offers several options, including the choice of integration algorithm through the `method` argument. For this problem, the LSODA algorithm was selected, as it is based on the Adams/BDF formulation and features automatic stiffness detection. The default Runge–Kutta method was not employed because preliminary results indicated insufficient accuracy in the concentration profiles of the second reactor.

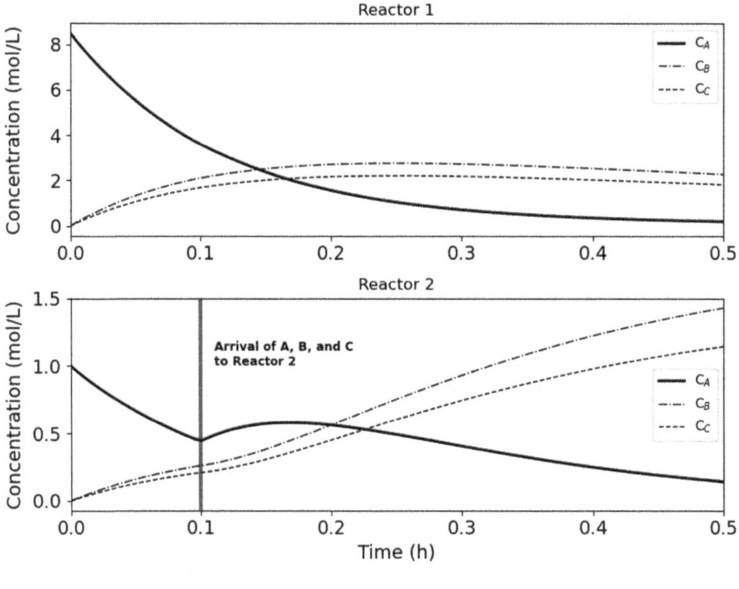

Fig. P3.2 Outlet concentration profiles for reactors 1 and 2

By integrating the differential system composed of the six equations, the following concentration profiles are obtained (Fig. P3.2).

Note that, in the second reactor, there is a break in the decreasing trend of compound A, because starting at $D = 0.1$ h, the excess of A from the first reactor begins to reach the second one.

6.4 Problem 4 Oscillating Tanks

The following diagram depicts two water tanks connected by a pipeline of diameter D_t (Fig. P4.1).

The state variables that describe the system are the liquid height of the first tank, $h_1(t)$; the liquid height of the second tank, $h_2(t)$; and the fluid velocity in the pipeline connecting the two tanks, $v_1(t)$. Using two mass balances and a momentum balance, the state equations are given by:

$$\frac{dh_1}{dt} = \frac{F_0 - F_1}{A_1} \tag{P4.1}$$

$$\frac{dh_2}{dt} = \frac{F_1 - F_2}{A_2} \tag{P4.2}$$

$$\frac{dv_1}{dt} = \frac{g \cdot (h_1 - h_2)}{L} - f \cdot \frac{v_1 \cdot |v_1|}{2 \cdot D_t} \tag{P4.3}$$

where

h_1	liquid height in tank 1
h_2	liquid height in tank 2
v_1	liquid velocity in the connecting pipeline
A_i	cross-sectional area of tank i
A_t	cross-sectional area of the pipeline connecting the two tanks

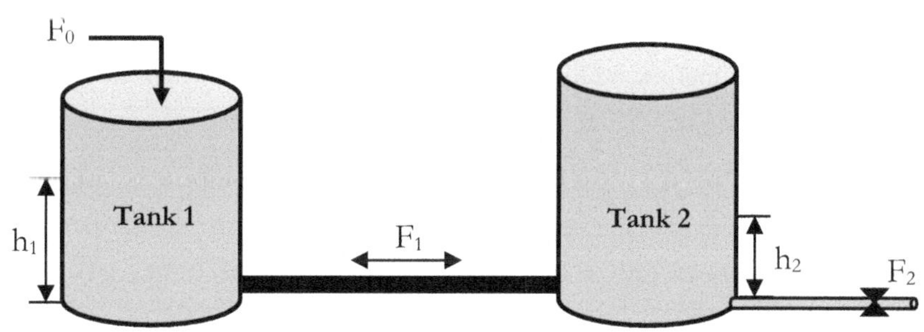

Fig. P4.1 Configuration of two connected tanks

$F_1 = v_1\,A_t$ liquid flow rate between the two tanks (which may occur in either direction)

F_0 and F_2 inlet and outlet flow rates of the system.

Consider the following constitutive equations for the friction factor in the pipeline (f), the flow area of the pipeline (A_t), the outlet flow rate from tank 2 (F_2), and the Reynolds number (Re), respectively:

$$f = \begin{cases} 64/\mathrm{Re} & \text{if Re} < 2000 \\ 0.18/\mathrm{Re}^{0.2} & \text{if Re} > 2000 \end{cases} \tag{P4.4}$$

where

$$A_t = \pi \cdot D_t^2/4$$

$$F_2 = c_v \cdot h_2^{0.5}$$

$$\mathrm{Re} = D \cdot \rho \cdot |v_1|/\mu$$

Data: The working liquid is water. The valve coefficient is $c_\mathrm{v} = 0.1\ \mathrm{m}^{2.5}/\mathrm{s}$. The diameters of the tanks are $D_1 = 2.5$ m and $D_2 = 3.0$ m, and the pipeline diameter is $D_\mathrm{t} = 0.3$ m. The inlet volumetric flow rate is $F_0 = 0.2\ \mathrm{m}^3/\mathrm{s}$. The gravitational acceleration is $g = 9.8\ \mathrm{m/s}^2$. The pipeline length is $L = 100$ m.

The fluid properties are a density $\rho = 1000\ \mathrm{kg/m}^3$ and a $\mu = 0.001\ \mathrm{kg\,m}^{-1}\,\mathrm{s}^{-1}$.
Initial conditions: $h_1(0) = 2$ m; $h_2(0) = 5$ m; $v_1(0) \approx 0$ m/s.
Based on the information given, answer the following questions:

(a) Plot the system dynamics $h_1(t)$, $h_2(t)$, $v_1(t)$ for the first 200 s.
(b) Now assume there are no inlet or outlet flows ($F_0 = F_2 = 0$). Plot the system response for the first 500 s. How can the physical behavior of the system be explained?
(c) Assume again that there are no inlet or outlet flows ($F_0 = F_2 = 0$), and that the frictional force in the pipeline of length L is negligible. What happens to the system dynamics in this case? Plot the system response.
(d) Finally, assume that the wall height of the first tank is 4 m. Plot the system dynamics for the first 500 s, assuming negligible friction factor and $F_0 = F_2 = 0$.

Solution

By directly integrating the differential equations, we obtain the following plots:

Figure P4.2 indicates that the system rapidly approaches steady state. In addition, the fluid velocity in the pipeline attains both positive and negative values, reflecting changes in the flow direction. An important feature of this problem is that, if the

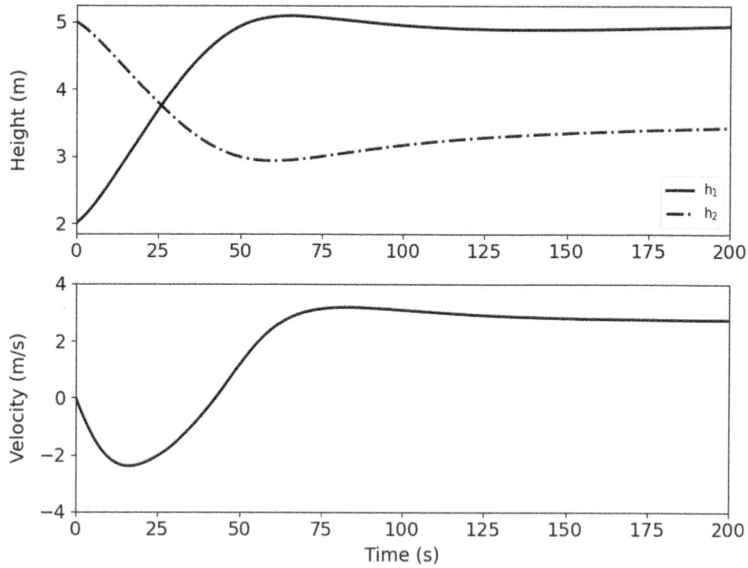

Fig. P4.2 Height and velocity response for part (a)

initial pipeline velocity were set exactly to zero, the numerical integration would fail to converge. Therefore, as specified in the problem statement, a value close to zero was used instead ($v(0) = 1 \times 10^{-8}$ m/s), which is effectively zero for practical purposes.

Figure P4.3 illustrates the case in which there is neither inlet flow ($F_0 = 0$) nor outlet flow ($F_2 = 0$). In this scenario, it can be observed that the oscillations of the tanks gradually decay due to energy dissipation in the pipeline that connects them. Once steady state is reached, we obtain $h_1 = h_2$. The reader is encouraged to compute this equilibrium height.[2]

Figure P4.4 shows the case in which there is no inlet flow ($F_0 = 0$) and no outlet flow ($F_2 = 0$), and the friction factor is negligible. In this situation, the maximum amplitude of the oscillations does not change over time due to the absence of energy dissipation. Under these conditions, the tanks behave as a perpetually oscillating system.

The code for part (b) is shown below:

[2] Answer = 3.77 m.

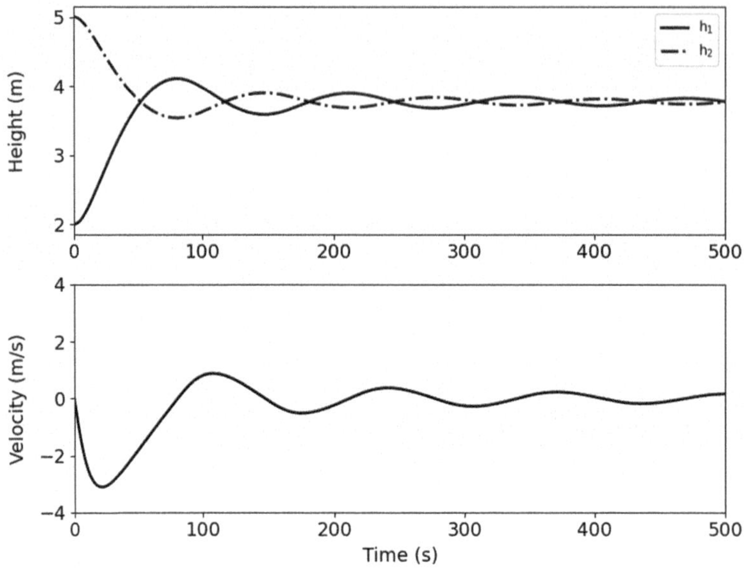

Fig. P4.3 Height and velocity response for part (b)

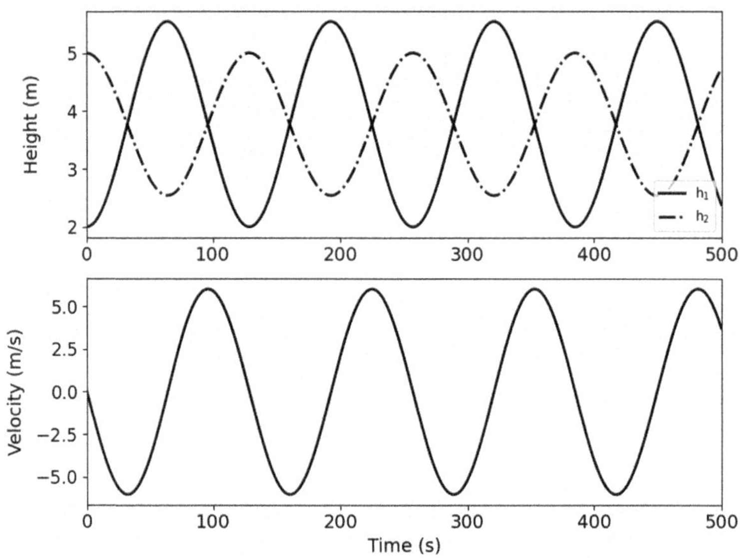

Fig. P4.4 Height and velocity response for part (c)

```
import numpy as np
import matplotlib.pyplot as plt
from scipy.integrate import solve_ivp

def est(t, x):
    # Parameters and constants
    cv = 0.1; D1 = 2.5; D2 = 3.0
    Dt = 0.3; F0 = 0.2*0; g = 9.8
    L = 100; mu = 0.001; ro = 1000

    # Constitutive relations
    A1 = np.pi*(D1**2)/4
    A2 = np.pi*(D2**2)/4
    At = np.pi*(Dt**2)/4
    Re = Dt*ro*np.abs(x[2])/mu
    F2 = cv*np.sqrt(np.abs(x[1]))*0

    # Friction factor
    if Re <= 2000:
        f = 64/Re
    else:
        f = 0.18/Re**0.2

    # Differential equations
    dx1 = (F0 - x[2]*At)/A1
    dx2 = (x[2]*At - F2)/A2
    dx3 = g/L*(x[0] - x[1]) - f*x[2]*np.abs(x[2])/(2*Dt)
    return [dx1, dx2, dx3]

# Initial conditions and integration time
x0 = [2, 5, 1e-8]
t_span = (0, 500)
t = np.linspace(t_span[0], t_span[1], 500)

# Integrate the differential equations
sol = solve_ivp(est, t_span, x0, t_eval=t, method='RK45')

# Extract results
h1, h2, v1 = sol.y

# Plot results
plt.figure(figsize=(8, 6))

plt.subplot(2, 1, 1)
plt.plot(t, h1, 'k', lw=2)
plt.plot(t, h2, 'k-.', lw=2)
plt.ylabel('Height (m)', fontsize=14, labelpad=20)
plt.legend(['h$_1$', 'h$_2$'], loc='best')
plt.xlim(0, 500)
plt.tick_params(axis='both', labelsize=14)

plt.subplot(2, 1, 2)
plt.plot(t, v1, 'k', lw=2)
plt.xlabel('Time (s)', fontsize=14)
plt.ylabel('Velocity (m/s)', fontsize=14, labelpad=8)
plt.xlim(0, 500)
plt.ylim(-4, 4)
plt.tick_params(axis='both', labelsize=14)

plt.tight_layout()
plt.show()
```

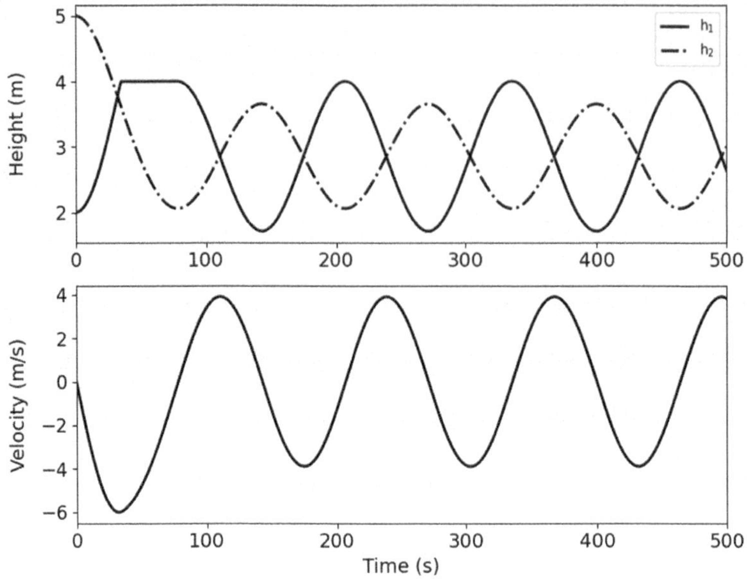

Fig. P4.5 Height and velocity response for part (d)

Case (d) is the most interesting to analyze, as it introduces the possibility of overflow in the smaller tank (tank 1). In this situation, the liquid level in tank 1 increases until it reaches the maximum allowable height of 4 m. The level then remains constant until the velocity in the connecting pipeline changes sign, indicating a reversal in the flow direction. This behavior occurs because the hydrostatic pressure difference between the tanks must reverse before net outflow from tank 1 can take place. Only after this reversal can the liquid level begin to decrease.

Figure P4.5 illustrates the system dynamics under the conditions described in part (d). In this case, the fluid in the first tank reaches the maximum height after 35 s. From that point onward, the system begins to lose water until the fluid velocity changes sign ($t = 78$ s).

The code for part (d) is shown below:

```
import numpy as np
import matplotlib.pyplot as plt
from scipy.integrate import solve_ivp

def est(t, x):
    # Parameters and constants
    cv = 0.1; D1 = 2.5; D2 = 3.0
    Dt = 0.3; F0 = 0; F2 = 0
    g = 9.8; h1max = 4.0; L = 100
    mu = 0.001; ro = 1000

    # Constitutive relations
    A1 = np.pi*(D1**2)/4
    A2 = np.pi*(D2**2)/4
    At = np.pi*(Dt**2)/4
    Re = Dt*ro*np.abs(x[2])/mu

    # Differential equations with tank overflow logic
    if x[0] >= h1max and x[2] < 0:
        dx1 = 0    # Tank 1 is full and cannot accept more liquid
    else:
        dx1 = (F0 - x[2]*At)/A1

    dx2 = (x[2]*At - F2)/A2
    dx3 = g/L*(x[0] - x[1])    # No friction term (negligible)
    return [dx1, dx2, dx3]

# Initial conditions and integration time
x0 = [2, 5, 1e-8]
t_span = (0, 500)
t = np.linspace(t_span[0], t_span[1], 1000)

# Integrate the differential equations
sol = solve_ivp(est, t_span, x0, t_eval=t, atol=1e-8, rtol=1e-8, method='RK45')

# Extract results
h1, h2, v1 = sol.y
datos = np.column_stack((t, sol.y.T))
np.savetxt('data_4d.txt', datos, delimiter=' ')

# Plot results
plt.figure(figsize=(8, 6))

plt.subplot(2, 1, 1)
plt.plot(t, h1, 'k', lw=2)
plt.plot(t, h2, 'k-.', lw=2)
plt.ylabel('Height (m)', fontsize=14, labelpad=20)
plt.legend(['h$_1$', 'h$_2$'], loc='upper right')
plt.xlim(0, 500)
plt.tick_params(axis='both', labelsize=14)

plt.subplot(2, 1, 2)
plt.plot(t, v1, 'k', lw=2)
plt.xlabel('Time (s)', fontsize=14)
plt.ylabel('Velocity (m/s)', fontsize=14, labelpad=8)
plt.xlim(0, 500)
plt.tick_params(axis='both', labelsize=14)

plt.tight_layout()
plt.show()
```

Case (d) requires the introduction of a special condition in the differential equation governing the liquid height in tank 1. Specifically, the model must account for overflow behavior. When the liquid level reaches or exceeds the maximum allowable height while the inlet velocity is negative (i.e., fluid is entering the tank), the tank is assumed to be full, and the time derivative of the liquid height is set equal to zero. This condition represents overflow and is enforced until the velocity changes sign, indicating that the tank is draining once again.

The behavior described above was incorporated into the differential equations solely using an `if-else` conditional. Another important aspect is that very small absolute and relative tolerances (1×10^{-8}) were used to ensure sufficient numerical precision, allowing the solver to accurately detect the moment at which the fluid height exceeds its maximum permitted value.

6.5 Problem 5 Parameter Estimation: Arrhenius Equation and Substrate Inhibition

(a) The Arrhenius expression is widely used to determine kinetic rate constants as a function of temperature:

$$k = A \cdot e^{-\frac{E}{R \cdot T}}, \tag{P5.1}$$

where $R = 1.987$ cal/gmol K is the ideal gas constant, and E is the activation energy of the reaction.

Table P5.1 shows the kinetic rate constants (k) as a function of temperature for a first-order reaction:

Based on this information:

- Determine the constants A and E of the Arrhenius model (including their corresponding units).
- Plot the experimental data versus the kinetic model (Eq. (P5.1)).
- Compute the squared error between the kinetic model and the data provided in Table P5.1.

(b) Under certain conditions, substrate inhibition models have been used to describe the specific growth rate (μ) of some microorganisms, such as the one presented below:

$$\mu = \frac{\mu_{\max} \cdot x}{k_m + x + k_1 \cdot x^2}, \tag{P5.2}$$

where μ max, k_1 y k_m are parameters, and x is the substrate concentration. Equation (P5.2) can also be expressed as:

$$\frac{1}{\mu} = \left(\frac{k_1}{\mu_{\max}}\right) \cdot x + \left(\frac{k_m}{\mu_{\max}}\right) \cdot \frac{1}{x} + \frac{1}{\mu_{\max}} \tag{P5.3}$$

Table P5.1 Reaction rate constants as a function of temperature

T (K)	300	310	320	330	340	350	360	370	380
k ($\min^{-1}$)	0.0014	0.0026	0.0047	0.0083	0.014	0.023	0.038	0.059	0.09

On the other hand, experiments carried out in a batch reactor yielded in Table P5.2.

Using this information:

- Determine the parameters k_1, k_m y μ_{max} (with their corresponding units) for the substrate inhibition model.
- Compare the model and the experiments by plotting μ versus x.
- Compute the squared error between the kinetic model and the experimental data.

Solution

(a) This problem can be solved in several ways. The first approach—perhaps the most straightforward for this type of situation—takes advantage of the structure of the Arrhenius equation in the following manner:

$$\ln(k) = \left(-\frac{E}{R}\right) \cdot \frac{1}{T} + \ln(A) \tag{P5.4}$$

Equation (P5.4) is identical to Eq. (P5.1), the only difference being that the natural logarithm has been applied. As can be seen, Eq. (P5.4) has the form of a straight line. In this case, the line has slope $-E/R$ and intercept $\ln(A)$. Given this, a linear regression can be performed using the transformed experimental data, that is, $1/T$ and $\ln(k)$. As a first strategy, we will use the `polyfit` function from the NumPy library.

Figure P5.1 shows the results obtained by applying `polyfit`. Figure P5.1a presents the linear model fitted to the transformed experimental data. Figure P5.1b displays the fitted Arrhenius model compared with the original experimental data. In both cases, the quality of the fit is excellent.

An alternative approach to solving this problem is to avoid linearization and instead employ the `curve_fit` function from the SciPy library. This routine performs nonlinear regression by estimating the model parameters through minimization of the sum of squared errors between the experimental data and the model predictions. By enabling the `full_output` option, additional information regarding the solution procedure can be obtained.

Table P5.3 summarizes the results obtained from both methods:

Table P5.3 shows that, based on the squared-error values, both methods perform very well. However, it is important to note that the `curve_fit` function requires an initial guess, which can introduce risk: a poor starting point may prevent the algorithm from converging. For this reason, it is advisable to test multiple initial guesses and draw on expert knowledge regarding the expected magnitude of the parameters.

Table P5.2 Growth rate as a function of substrate concentration

x (g/L)	0.10	0.15	0.25	0.50	0.75	1.00	1.50	3.00
μ (h^{-1})	0.24	0.27	0.34	0.35	0.35	0.34	0.33	0.22

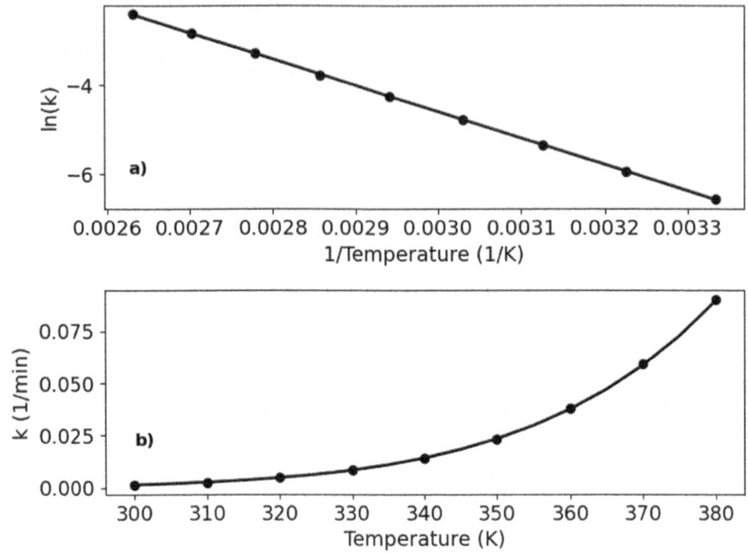

Fig. P5.1 **a** Comparison between the linear model and the experimental data. **b** Comparison between the Arrhenius model and the experimental data

Table P5.3 Comparison between the linear model and the `curve_fit` function

	Linear model	`Curve_fit`
E (cal/gmol)	11,824	11,882
A (1/min)	565,798	615,365
Sum of squared errors	9.03×10^{-7}	3.65×10^{-7}

(b) The procedure used in part (a) can be applied here as well; that is, a transformation may be sought to linearize the model. In this case, the inverse of the function is given by:

$$\frac{1}{\mu} = \left(\frac{k_1}{\mu_{\max}}\right) \cdot x + \left(\frac{k_m}{\mu_{\max}}\right) \cdot \frac{1}{x} + \frac{1}{\mu_{\max}} \tag{P5.5}$$

Equation (P5.5) is of the form:

$$y = a \cdot x_1 + b \cdot x_2 + c \tag{P5.6}$$

where

$$a = k_1/\mu_{\max}, \quad b = k_m/\mu_{\max}, \quad c = 1/\mu_{\max}, \quad x_2 = 1/x$$

This equation can be solved by applying the analytical solution for multiple linear regression, shown on the right-hand side of Eq. (P5.7). The analytical expression is:

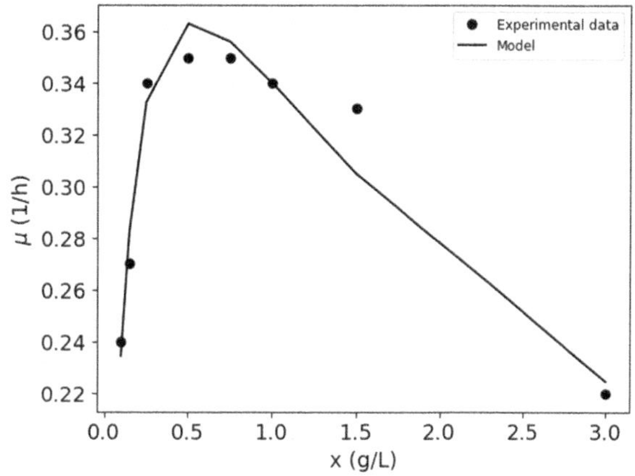

Fig. P5.2 Comparison between experimental data and model fitted using multilinear regression

$$\beta = \left(X^T \cdot X\right)^{-1} \cdot X^T \cdot Y, \tag{P5.7}$$

where β is the parameter vector obtained through regression, Y is the vector of transformed data $1/\mu$, and X is the matrix containing the regressors. Thus:

$$\beta = \begin{bmatrix} k_1/\mu_{\max} \\ k_m/\mu_{\max} \\ 1/\mu_{\max} \end{bmatrix}; \quad Y = \begin{bmatrix} 1/\mu(1) \\ 1/\mu(2) \\ \vdots \\ 1/\mu(8) \end{bmatrix}; \quad X = \begin{bmatrix} x(1) & 1/x(1) & 1 \\ x(2) & 1/x(2) & 1 \\ \vdots & \vdots & \vdots \\ x(8) & 1/x(8) & 1 \end{bmatrix}$$

To solve this regression, we will use the Statsmodels[3] library, which provides extensive diagnostic information about fitness. This information can be inspected through the summary table obtained with `report.summary`. The reader is encouraged to review the report by running the accompanying Python program.

Figure P5.2 shows the result of the multilinear fit compared with the experimental data.

As in part (a), the `curve_fit` function can be used directly, without invoking Eq. (P5.3). The following table compares the results obtained using both solution methods:

Table P5.4 shows that the squared errors are almost identical for both methods. Moreover, the differences in the estimated parameter values result in concentration profiles that are practically indistinguishable when plotted.

The Python program developed for part (a) is presented below:

[3] https://www.statsmodels.org/stable/index.html.

Table P5.4 Comparison between the multilinear model and the `curve_fit` function

	Multilinear model	`Curve_fit`
μ max (1/h)	0.532	0.513
k_1 (L/g)	0.443	0.390
k_m (g/L)	0.123	0.115
Sum of squared errors	1.1×10^{-3}	9.9×10^{-4}

```
import numpy as np
import matplotlib.pyplot as plt
from scipy.optimize import curve_fit

# Constants and data
R = 1.987  # cal/gmol*K
k_exp = np.array([0.0014, 0.0026, 0.0047, 0.0083, 0.014, 0.023, 0.038, 0.059, 0.09])
T = np.arange(300, 381, 10)
invT = 1.0 / T
y = np.log(k_exp)

# First strategy: first-order regression (ln(k) = A + B * (1/T))
coef = np.polyfit(invT, y, 1)

# Graphical comparison between model and experimental data (error)
plt.figure(figsize=(8, 6))

plt.subplot(2, 1, 1)
plt.plot(invT, y, 'ko', invT, np.polyval(coef, invT), 'k', lw=2)
plt.xlabel('1/Temperature (1/K)', fontsize=14)
plt.ylabel('ln(k)', fontsize=14)
plt.text(0.002625, -6, "a)", fontsize=12, fontweight='bold')
plt.subplots_adjust(hspace=0.4)
plt.tick_params(axis='both', labelsize=14)

plt.subplot(2, 1, 2)
T1 = np.arange(300, 381, 5)
k_model = np.exp(coef[1]) * np.exp(coef[0] / T1)
plt.plot(T, k_exp, 'ko', T1, k_model, 'k', linewidth=2)
plt.xlabel('Temperature (K)', fontsize=14)
plt.ylabel('k (1/min)', fontsize=14)
plt.text(300, 0.02, "b)", fontsize=12, fontweight='bold')
plt.tick_params(axis='both', labelsize=14)
plt.show()

# Numerical results
A = np.exp(coef[1])
E = -coef[0] * R
print('- Linear fit (polyfit):')
print(f'A = {round(A,1)} 1/min')
print(f'E = {round(E,1)} cal/gmol')

# Sum of squared errors (SSE) between model and data
k_model = A * np.exp(coef[0] * invT)
error = np.sum((k_exp - k_model)**2)
print(f'Error (SSE) = {round(error, 9)}')

# Second strategy: nonlinear least-squares fit (curve_fit)
print('\n- Nonlinear fit (curve_fit):')
```

```
# Arrhenius model
def modelo(xdata, A, E_R):
    return A * np.exp(-E_R / (R * xdata))

# Initial guess and execution
initial_guess = [500000, 1e4]
params, pcov, _, mesg, _ = curve_fit(
    modelo,
    xdata=T,
    ydata=k_exp,
    p0=initial_guess,
    full_output=True
)
print("Convergence message:", mesg)

# SSE for the nonlinear fit
model_predictions = modelo(T, *params)
sse = np.sum((k_exp - model_predictions)**2)

print(f'A = {round(params[0],1)} 1/min')
print(f'E = {round(params[1],1)} cal/gmol')
print("Error (SSE) =", round(sse, 9))
```

The Python code developed for part (b) is presented below:

```
import numpy as np
import matplotlib.pyplot as plt
import statsmodels.api as sm
from scipy.optimize import curve_fit

# Experimental data
xdata = np.array([0.1, 0.15, 0.25, 0.5, 0.75, 1., 1.5, 3.])
mu = np.array([0.24, 0.27, 0.34, 0.35, 0.35, 0.34, 0.33, 0.22])

# STATSMODELS: Multivariable regression
Y = 1/mu
X = sm.add_constant(np.column_stack((xdata, 1/xdata, np.ones(8))))
model = sm.OLS(Y, X)
results = model.fit()

print('- Results using linear regression (Statsmodels):')
mu_max = 1/results.params[2]
k1 = mu_max * results.params[0]
km = mu_max * results.params[1]
print(f'mu_max: {round(mu_max,3)}')
print(f'k1: {round(k1,3)}')
print(f'km: {round(km,3)}')

# Sum of squared errors (SSE) between model and data
predicciones = 1 / results.predict(X)
sse = np.sum((mu - predicciones)**2)
print(f"Error (SSE) = {round(sse,5)}\n")

# Regression summary
print(results.summary())

# Plotting experimental data versus model
plt.plot(xdata, mu, 'ko', label='Experimental data')
plt.plot(xdata, predicciones, 'k-', label='Model')
plt.xlabel('x (g/L)', fontsize=14)
plt.ylabel('mu (1/h)', fontsize=14)
plt.tick_params(axis='both', labelsize=14)
plt.legend(fontsize=9)
plt.show()

# CURVE_FIT: nonlinear parameter estimation
def model_func(x, mu_max, km, kl):
    return (mu_max * x) / (km + x + kl * x**2)
```

```
initial_guess = [1.0, 1.0, 1.0]
params, _, _, mesg, _ = curve_fit(
    model_func,
    xdata=xdata,
    ydata=mu,
    p0=initial_guess,
    full_output=True
)
print("- Convergence message:", mesg)

print("- Results using nonlinear regression (curve_fit):")
mu_max = params[0]
km = params[1]
kl = params[2]
print(f"mu_max = {round(mu_max,3)}")
print(f"kl = {round(kl,3)}")
print(f"km = {round(km,3)}")

# SSE for nonlinear model
model_predic = model_func(xdata, *params)
sse = np.sum((mu - model_predic)**2)
print(f"Error (SSE) = {round(sse,5)}")
```

Later, we will examine how to estimate parameters of differential equations from experimental data (Problems 9 and 10).

6.6 Problem 6 Parameter Estimation and Confidence Intervals: Substrate Inhibition in Biological Systems

Under specific conditions, substrate inhibition models have been used to describe the specific growth rate (μ) of certain microorganisms, as illustrated below:

$$\mu = \frac{\mu_{\max} \cdot x}{k_m + x + k_1 \cdot x^2}, \tag{P6.1}$$

where μ max k_1 y k_m are parameters, and x denotes the substrate concentration. Furthermore, experiments conducted in a batch reactor yielded the following relationship between biomass and specific growth rate (Table P6.1).

Using the information provided, determine the parameters k_1, k_m y μ max together with their corresponding 95% confidence intervals. Compare the model with the experimental data by plotting μ versus x, including 95% confidence intervals for the model predictions. Discuss your results, comparing them with those obtained in Problem 5.

Table P6.1 Specific growth rate as a function of substrate concentration

x (g/L)	0.10	0.15	0.25	0.50	0.75	1.00	1.50	3.00
μ (h^{-1})	0.24	0.27	0.34	0.35	0.35	0.34	0.33	0.22

Solution

To estimate the confidence intervals of the parameters, the `curve_fit` function from the SciPy library is first employed. In addition to providing parameter estimates, this routine yields the covariance matrix of the estimated parameters through its second output, `pcov`. The standard errors of the parameters are obtained by taking the square roots of the diagonal elements of this matrix, from which the corresponding confidence intervals can be constructed.

The standard error, when multiplied by the Student's t statistic for a given significance level α (e.g., 0.05), yields the $(1 - \alpha)$ confidence intervals. The complete procedure is described step by step in the accompanying Python code.

The parameters together with their corresponding confidence intervals are presented in Table P6.2:

To plot the model fit together with its 95% confidence intervals, the Jacobian of the model with respect to its three parameters μ max, k_1, and k_m must be evaluated. This is accomplished by defining the Jacobian matrix containing the corresponding partial derivatives. Once obtained, the Jacobian and the parameter covariance matrix ($pcov$) are used to estimate the propagation of parameter uncertainty into the model prediction, yielding the standard deviation of the model output, $\sigma\mu$.

In mathematical terms:

$$\sigma_{\mu}^2 = \text{diag}\left(J \cdot p\,\text{cov} \cdot J^T\right) \tag{P6.2}$$

where

σ_{μ}^2 variance vector of the model predictions (μ). Since we have eight experimental data points, this vector has dimension 8×1.
J Jacobian matrix of the model with respect to the three parameters (8×3 matrix).
J^{T} $p$cov covariance matrix of the estimated parameters (3×3 matrix) transpose of the Jacobian matrix (3×8 matrix).

By multiplying these three matrices (Eq. (P6.2)), we obtain an 8×8 matrix. Extracting its diagonal yields an 8×1 vector, which represents the variance of the predicted values at each experimental point due to parameter uncertainty. To compute the confidence interval, we must obtain the standard deviation (by taking the square root of the variance vector) and multiply it by the Student's t statistic corresponding to the selected significance level α.

Table P6.2 Parameters and 95% confidence intervals

Parameters	Mean ± Confidence interval
μ max (1/h)	0.513 ± 0.101
k_1 (L/g)	0.390 ± 0.215
k_m (g/L)	0.115 ± 0.059
Sum of squared errors	0.00099

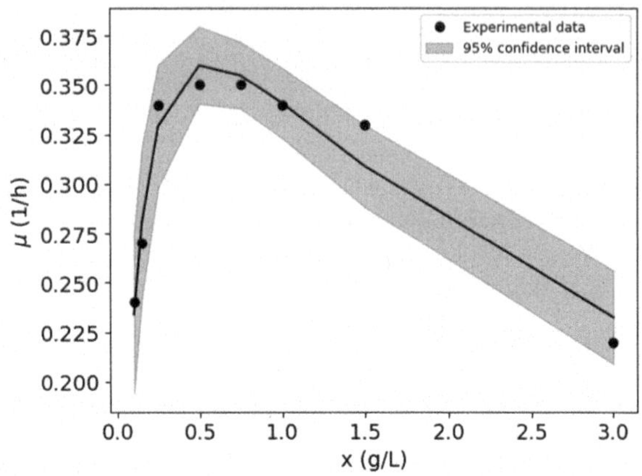

Fig. P6.1 Experimental data and model prediction (solid line) with 95% confidence intervals (shaded region)

Thus, the lower and upper confidence limits for the predicted values of μ (μ pred) are computed as:

$$\mu_{\text{sup}} = \mu_{\text{pred}} + t_{\alpha=0.05} \cdot \sigma_{\mu} \tag{P6.3}$$

$$\mu_{\text{inf}} = \mu_{\text{pred}} - t_{\alpha=0.05} \cdot \sigma_{\mu} \tag{P6.4}$$

In this problem, the Jacobian was evaluated explicitly, step by step. The procedure used to compute and plot the confidence intervals is general and can be extended to more complex models. In such cases, it is often convenient to automate the evaluation of the Jacobian using Python libraries such as Autograd or JAX, which are available through the public PyPI repository.

Shown below are the fitted model prediction μ pred, the corresponding 95% confidence bounds, μ sup and μ inf, and the experimental data (Fig. P6.1).

The Python code for this problem is shown below:

```
import numpy as np
import matplotlib.pyplot as plt
from scipy.optimize import curve_fit
from scipy.stats import t

# Experimental data
xdata = np.array([0.1, 0.15, 0.25, 0.5, 0.75, 1.0, 1.5, 3.0])
mu = np.array([0.24, 0.27, 0.34, 0.35, 0.35, 0.34, 0.33, 0.22])

# CURVEFIT: nonlinear parameter estimation
def model_func(x, mu_max, km, kl):
    return mu_max*x/(km + x + kl*x**2)

initial_guess = [1.0, 1.0, 1.0]
params, pcov,_,mesg,_ = curve_fit(model_func, xdata=xdata, ydata=mu, p0=initial_guess,
full_output=True)
print("- Convergence message:", mesg)

# Calculation of confidence intervals
n = len(mu)  # number of experimental data points
p = len(params)  # number of parameters

# Degrees of freedom
df = max(0, n - p)

# Confidence interval and Student's t distribution for a given alpha
alpha = 0.05
t_value = t.ppf(1 - alpha/2, df)  # t-value for a given alpha and degrees of freedom

# Calculation of standard error and confidence intervals
perr = np.sqrt(np.diag(pcov))
ci = t_value * perr

print("- Results obtained using nonlinear regression (curve_fit):")
mu_max = params[0]
km = params[1]
kl = params[2]
print(f"mu_max = {mu_max:.3f} ± {ci[0]:.3f}")
print(f"km = {km:.3f} ± {ci[1]:.3f}")
print(f"kl = {kl:.3f} ± {ci[2]:.3f}")

# Calculation of the squared error between model and data (SSE)
mu_pred = model_func(xdata, *params)
sse = np.sum((mu - mu_pred)**2)
print(f"Error (SSE) = {round(sse,5)}")

# Plotting the 95% confidence intervals
x_range = xdata  # For a smoother curve, use: np.linspace(min(xdata), max(xdata), 400)
mu_pred = model_func(x_range, *params)

# Jacobian calculation for this problem
def jacobian(x, mu_max, km, kl):
    J = np.empty((len(x), len(params)))
    D = km + x + kl * x**2
    J[:, 0] = x/(D)                  # Partial derivative with respect to mu_max
    J[:, 1] = -(mu_max*x)/(D)**2     # Partial derivative with respect to km
    J[:, 2] = -(mu_max*x**2)/(D)**2  # Partial derivative with respect to kl
    return J

J = jacobian(x_range, *params)
sigma_mu = np.sqrt(np.sum((J @ pcov)*J, axis=1))  # Equivalent to np.sqrt(np.diag(J @
pcov @ J.T)) but faster
mu_upper = mu_pred + t_value * sigma_mu
mu_lower = mu_pred - t_value * sigma_mu

# Plotting the results
plt.plot(xdata, mu, 'ko', label='Experimental data')
plt.plot(xdata, mu_pred, 'k-')
plt.xlabel('x (g/L)', fontsize=14)
plt.ylabel('$\\mu$ (1/h)', fontsize=14)
plt.fill_between(x_range, mu_lower, mu_upper,
                 color='gray', alpha=0.35,
                 label='95% Confidence Interval')
plt.tick_params(axis='both', labelsize=14)
plt.legend(fontsize=9)

plt.show()
```

Very large parameter confidence intervals indicate potential identifiability issues. In such cases, additional experiments can be performed to increase the degrees of freedom, or the parameters may be re-estimated while fixing one of them at a known value. In the latter approach, the fixed parameter values may be obtained from the relevant literature or inferred from the physical characteristics of the system. Either strategy serves to reduce the uncertainty and narrow the resulting confidence intervals.

6.7 Problem 7 Continuous Cultivation Bioreactor: Monod and Substrate Inhibition Kinetics

Biochemical reactors, commonly referred to as bioreactors, are used to produce a wide range of intermediate and final products, including pharmaceuticals, foods, and beverages, and are also widely employed in waste treatment processes. The mathematical models used to describe bioreactors are closely analogous to those presented earlier for chemical reactors, as they are based on the same fundamental material balances.

In the simplest bioreactor models, two components are typically considered: biomass and substrate (or nutrient). The biomass consists of living cells that consume the substrate. Examples include liquid waste treatment, where biomass degrades organic matter, and wine production by fermentation, in which yeast consumes sugar and produces ethanol along with other minor compounds.

The nomenclature used in this case study is as Fig. P7.1.

Assuming a perfectly mixed reactor, the following equations represent the basic material balances for a continuous-culture bioreactor. Because the inlet and outlet flow rates are equal, the liquid volume remains constant:

Biomass Growth

The total accumulation of biomass per unit time is equal to the rate at which biomass enters the system minus the rate at which it leaves, plus the biomass generated inside the bioreactor per unit time.

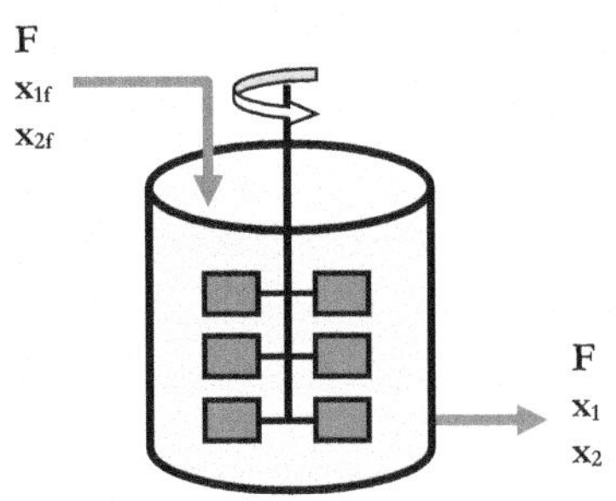

F : volumetric flow rate.
r_1 : cell generation rate.
r_2 : substrate consumption rate.
x_1 : biomass concentration.
x_2 : substrate concentration.
V : bioreactor volume.
μ : specific growth rate.

Fig. P7.1 Bioreactor configuration

$$\frac{d(V \cdot x_1)}{dt} = F \cdot x_{1f} - F \cdot x_1 + V \cdot r_1, \tag{P7.1}$$

where $r_1 = \mu \cdot x_1$

Substrate Consumption

In this case, the accumulation of nutrients per unit time is equal to the rate at which substrate enters the system minus the rate at which it leaves, and the amount of substrate consumed per unit time.

$$\frac{d(V \cdot x_2)}{dt} = F \cdot x_{2f} - F \cdot x_2 - V \cdot r_2 \tag{P7.2}$$

Furthermore, there exists a relationship between the cell generation rate and the substrate consumption rate. This relationship is known as the yield coefficient (Y) and is expressed as follows:

$$Y = \frac{\text{mass of cell produced}}{\text{mass of substrate consumed}} = \frac{r_1}{r_2} \tag{P7.3}$$

From the above definition, it follows that $r_2 = \mu \cdot X_1/Y$. For practical purposes, the yield coefficient Y is assumed to be constant. Furthermore, if the reactor volume remains constant, the dilution rate may be defined as $D = F/V$ (where D is known as the dilution rate or simply dilution). Substituting Y and D into Eqs. (P7.1) and (P7.2) yields the governing equations for biomass and substrate:

$$\frac{dx_1}{dt} = D \cdot \left(x_{1f} - x_1\right) + \mu \cdot x_1 \tag{P7.4}$$

$$\frac{dx_2}{dt} = D \cdot \left(x_{2f} - x_2\right) - \mu \cdot x_1/Y \tag{P7.5}$$

In general, the inlet stream is assumed to contain no biomass; that is, $x_{1f} = 0$. The resulting equations describing biomass growth (x_1) and nutrient consumption (x_2) are:

$$\frac{dx_1}{dt} = (\mu - D) \cdot x_1 \tag{P7.6}$$

$$\frac{dx_2}{dt} = D \cdot (x_{2f} - x_2) - \mu \cdot x_1/Y \tag{P7.7}$$

Cell Growth Kinetics

The most widely used constitutive relationship to describe the specific growth rate is the Monod equation:

$$\mu = \frac{\mu_{\max} \cdot x_2}{k_m + x_2} \quad \text{(P7.8)}$$

Equation (P7.8) relates biomass growth to the concentration of a limiting substrate. In this case, the growth rate decreases once the limiting substrate begins to be depleted. The term k_m corresponds to a saturation constant for the limiting substrate (x_2) and μ max denotes the maximum specific growth rate. On average, bacteria exhibit μ max values close to 0.9 1/h, yeasts around 0.45 1/h, and filamentous fungi approximately 0.25 1/h; nevertheless, μ max must be determined experimentally for each specific case (Bailey, 1986).

In some cases, a nutrient in the culture medium may inhibit growth, particularly when present at relatively high concentrations. This is likely to occur with the carbon and energy source, as it is typically the component present in the highest proportion. To model this effect, the following expression is commonly used:

$$\mu = \frac{\mu_{\max} \cdot x_2}{k_m + x_2 + k_1 \cdot x_2^2} \quad \text{(P7.9)}$$

The fact that a high substrate concentration may inhibit growth imposes a restriction on the allowable nutrient concentration and, consequently, sets an upper limit on the maximum final biomass concentration that can be achieved. Such considerations naturally suggest the formulation of an optimization problem aimed at maximizing the biomass concentration.

Based on the preceding discussion and assuming a bioreactor of constant volume:

(a) Compute the steady states of the system for $D = 0.15\ \text{h}^{-1}$, $D = 0.30\ \text{h}^{-1}$ y $D = 0.45\ \text{h}^{-1}$, using the initial condition $[x_{10}, x_{20}] = [1.0, 1.0]$ and $[x_{10}, x_{20}] = [0.75, 2.0]$. Compare the results for both Monod kinetics and substrate inhibition kinetics.

(b) Determine the total amount of biomass produced after 30 h of fermentation for $D = 0.30\ \text{h}^{-1}$ and $[x_{10}, x_{20}] = [0.75, 2.0]$, when glucose is used as the nutrient. Compare the Monod and substrate inhibition kinetics. Assume that the reactor volume is $V = 1$ L Table P7.1.

Solution

(a) In this case, an alternative approach for computing the steady state is illustrated, without resorting to the use of the `fsolve` function. The strategy consists of

Table P7.1 Relevant model parameters Bequette (1998)

	Monod	Substrate inhibition
μ max (1/h)	0.53	0.53
k_m (g/L)	0.12	0.12
k_1 (L/g)	–	0.45
Y	0.4	0.4
x_{2f} (g/L)	4.0	4.0

integrating the system of equations over a sufficiently long-time horizon. The key idea is to extract the final values of the state variables once the system has reached steady state. A graphical illustration of this approach is shown below:

Figure P7.2 shows that the system approaches steady state after a sufficiently long period of operation. As indicated, the steady-state values are obtained at approximately $t = 300$ h, corresponding to the final point of the simulation.

What advantages does this method offer compared with using the `fsolve` function? The main advantage is that, if a steady state exists, time integration will always allow it to be determined directly. In contrast, `fsolve` may experience convergence issues when dealing with highly nonlinear or stiff systems. In both approaches, however, multiple solutions or steady states may be obtained depending on the initial guess used to initialize the method.

Table P7.2 presents the different steady states for biomass and glucose for the initial conditions $[x_{10}, x_{20}] = [1.0, 1.0]$ and $[x_{10}, x_{20}] = [0.75, 2.0]$.

For cultures governed by Monod kinetics, the first observation is that the steady-state biomass concentration decreases slightly as the dilution rate increases. In contrast, the effect of dilution is more pronounced in the case of substrate inhibition, where the biomass may be completely washed out at $D = 30$ and $0.45\ \text{h}^{-1}$. This behavior, known as *washout*, occurs when the rate at which cells are removed from the bioreactor exceeds their net growth rate.

For completeness, a typical washout curve for substrate inhibition at $D = 0.45\ \text{h}^{-1}$, is shown in Fig. P7.3:

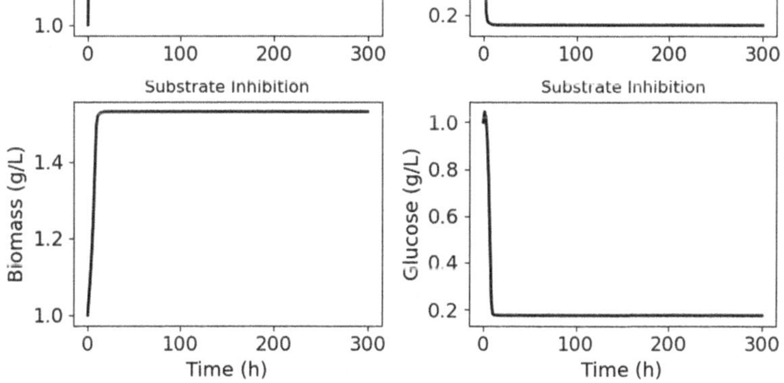

Fig. P7.2 Simulation for $D = 0.30\ \text{h}^{-1}$ and $[x_{10}, x_{20}] = [1.0, 1.0]$

Table P7.2 Steady-state values as a function of dilution rate and initial condition

	Monod kinetics		Substrate inhibition	
	[1.0, 1.0]	[0.75, 2.0]	[1.0, 1.0]	[0.75, 2.0]
D = 0.15 h^{-1}				
Biomass	1.58	1.58	1.58	1.58
Glucose	0.05	0.05	0.05	0.05
D = 0.30 h^{-1}				
Biomass	1.54	1.54	1.53	0.0
Glucose	0.16	0.16	0.17	4.0
D = 0.45 h^{-1}				
Biomass	1.33	1.33	0.0	0.0
Glucoss	0.67	0.67	4.0	4.0

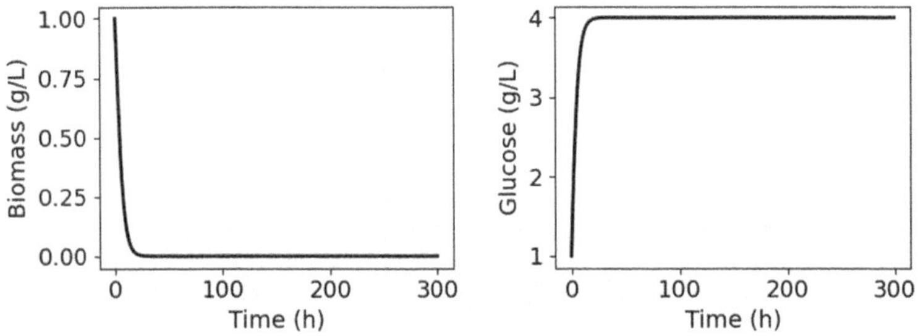

Fig. P7.3 Simulation of a washout event (substrate inhibition, $D = 0.45\ h^{-1}$)

(b) Strictly speaking, the total biomass produced must be obtained by integrating the biomass flow leaving the bioreactor over the time interval from 0 to 30 h and adding the biomass remaining in the reactor at the end of the fermentation, while subtracting the initial biomass present. Mathematically, the problem may be stated as follows:

$$\text{Total biomass} = \int_0^{30} F \cdot x_1 \cdot \mathrm{d}t + V \cdot (x_1(t = 30) - x_1(t = 0)). \qquad \text{(P7.10)}$$

Figure P7.4 shows the total amount of biomass produced after 30 h of fermentation. Under the conditions considered, the model based on Monod kinetics yields the highest biomass production. This result is expected, since the steady-state biomass concentration predicted by Monod kinetics is significantly higher than that obtained in the presence of substrate inhibition (see Table P7.3). Moreover, Monod kinetics does not account for inhibitory effects on growth. In practice, however, biological systems invariably exhibit some form of limitation.

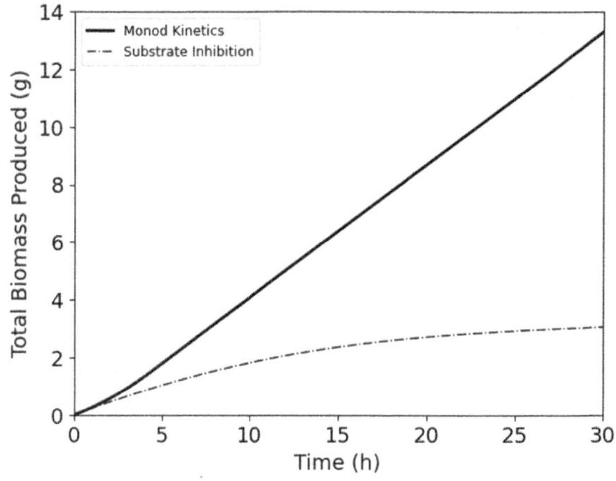

Fig. P7.4 Biomass that has left the bioreactor ($D = 0.30\ h^{-1}$)

Table P7.3 Total biomass produced after 30 h of fermentation

	Monod kinetics	Substrate inhibition
Total biomass produced (g)	14.09	2.37

The results of biomass production for both kinetic models are summarized in Table P7.3.

The Python code used to solve part (b) of this problem is presented below:

```
import numpy as np
import matplotlib.pyplot as plt
from scipy.integrate import solve_ivp

# Bioreactor model
def bioreact(t, x):
    # Parameters
    mumax = 0.53  # (1/h)
    km = 0.12     # (g/L)
    k1 = 0.45     # (L/g)
    Y = 0.4       #
    x2f = 4.0     # (g/L)
    D = 0.30       # (1/h)
    V = 1.0       # (L)

    # Specific growth rates
    mu1 = mumax*x[1]/(km + x[1])
    mu2 = mumax*x[3]/(km + x[3] + k1*x[3]**2)
    # State variables: Monod kinetics
    dx1 = (mu1 - D)*x[0]
```

```
    dx2 = D*(x2f - x[1]) - mu1*x[0]/Y
    # State variables: substrate inhibition
    dx3 = (mu2 - D)*x[2]
    dx4 = D*(x2f - x[3]) - mu2*x[2]/Y
    # Biomass leaving the bioreactor
    dx5 = D*V*x[0]
    dx6 = D*V*x[2]
    return [dx1, dx2, dx3, dx4, dx5, dx6]

# Initial conditions for (Monod and inhibition): biomass, glucose, and biomass that has
left the bioreactor
X0 = [0.75, 2.0, 0.75, 2.0, 0.0, 0.0]

# Integration times
t_span = (0, 30)
t_eval = np.linspace(t_span[0], t_span[1], 200)  # Evaluation points

# Solving the system of equations
sol = solve_ivp(bioreact, t_span, X0, t_eval=t_eval, atol=1e-8, rtol=1e-8, method='RK45')

# Extracting results
t = sol.t; y = sol.y

# Plotting results
plt.figure(1, figsize=(8,6))

plt.subplot(2, 2, 1)
plt.plot(t, y[0], 'k', lw=2)
plt.ylabel('Biomass (g/L)', fontsize=14)
plt.tick_params(axis='both', labelsize=14)
plt.title('Monod Kinetics', fontsize=12)

plt.subplot(2, 2, 2)
plt.plot(t, y[1], 'k', lw=2)
plt.ylabel('Glucose (g/L)', fontsize=14)
plt.tick_params(axis='both', labelsize=14)
plt.title('Monod Kinetics', fontsize=12)

plt.subplot(2, 2, 3)
plt.plot(t, y[2], 'k', lw=2)
plt.xlabel('Time (h)', fontsize=14)
plt.ylabel('Biomass (g/L)', fontsize=14)
plt.title('Substrate Inhibition', fontsize=12)
plt.tick_params(axis='both', labelsize=14)

plt.subplot(2, 2, 4)
plt.plot(t, y[3], 'k', lw=2)
plt.xlabel('Time (h)', fontsize=14)
plt.ylabel('Glucose (g/L)', fontsize=14)
plt.title('Substrate Inhibition', fontsize=12)
plt.tick_params(axis='both', labelsize=14)
plt.tight_layout()

plt.figure(2)
plt.plot(t, y[4], 'k', lw=2, label='Monod Kinetics')
plt.plot(t, y[5], 'k-.', lw=1, label='Substrate Inhibition')
plt.xlabel('Time (h)', fontsize=14)
plt.xlim(0, 30)
plt.ylabel('Total Biomass Produced (g)', fontsize=14)
plt.ylim(0, 14)
plt.legend(fontsize=9)
plt.tick_params(axis='both', labelsize=14)

plt.show()

# Final value of the state variables
ultimo_valor = y[:, -1]
print("Final values at the end of fermentation:\n", ultimo_valor)

# Calculation of total biomass produced
biomasa_producida_monod = y[4, -1] + 1*(y[0, -1] - X0[0])
print(f'\nMonod, total biomass =', round(biomasa_producida_monod,2))
biomasa_producida_inhibicion = y[5, -1] + 1*(y[2, -1] - X0[2])
print(f'Substrate inhibition, total biomass =', round(biomasa_producida_inhibicion,2))
```

6.8 Problem 8 Parameter Estimation: Ordinary Differential Equations (ODEs)

Unlike Problem 5, the present problem requires estimation of the yield coefficient $Y_{x/s}$ and the maintenance coefficient m_s appearing in an ordinary differential equation (ODE). This equation is part of a larger model comprising eight ODEs that describe the cultivation of the filamentous fungus *Gibberella fujikuroi* on a solid substrate at laboratory scale (Gelmi et al., 2002).

A simplified version of the model, consisting of four differential equations, is presented below:

Active Biomass

The evolution of active biomass is defined by a balance that includes a specific growth rate (μ) and a fungal death term; the latter is assumed to be first order.

$$\frac{\mathrm{d}X}{\mathrm{d}t} = \mu \cdot X - k_d \cdot X \tag{P8.1}$$

Urea Degradation

The degraded urea is primarily utilized for the formation of structural proteins, which constitute part of the microorganism's cell walls. The model assumes that the limiting nitrogen source, urea (U), is decomposed into an intermediate nitrogen species (N_I; for example, ammonium, which is subsequently incorporated into the biosynthesis of nitrogen-containing compounds, following zero-order kinetics. Once the urea in the medium is exhausted, only the consumption of N_I continues.

$$\frac{\mathrm{d}U}{\mathrm{d}t} = -k \tag{P8.2}$$

$$\frac{\mathrm{d}N_I}{\mathrm{d}t} = 0.47 \cdot k - \mu \cdot \frac{X}{Y_{X/N_I}} \tag{P8.3}$$

In Eq. (P8.3), the factor 0.47 corresponds to the mass fraction of nitrogen in urea.

Degradation of the Carbon Source

The first term in the carbon-degradation kinetics accounts for fungal growth, providing the elements required for biomass synthesis as well as the energy necessary for this process. The second term corresponds to maintenance, which represents the energy required to keep the microorganism alive and supplies the materials needed for the formation of extracellular products. The parameters $Y_{X/S}$ y m_s are the quantities to be estimated in this problem.

$$\frac{\mathrm{d}S}{\mathrm{d}t} = -\frac{\mu \cdot X}{Y_{X/S}} - m_s \cdot X \tag{P8.4}$$

Constitutive Equation: Specific Growth Rate

In this case, it is necessary to model biomass evolution under nutrient-limited conditions. Accordingly, the dependence of μ on nitrogen was modeled using a Monod-type relationship.

$$\mu = \frac{\mu_{\max} \cdot N_I}{(N_I + k_n)} \tag{P8.5}$$

The relevant model parameters, together with the ideal operating conditions, are summarized in Table P8.1.

Table P8.2 presents the experimental measurements of the carbon source.

Solution

Two solution strategies are employed in this problem. The first uses the `least_squares` optimization routine from the SciPy library. This routine is well suited for problems in which a residual function is explicitly minimized and allows the specification of parameter bounds, among other features, thereby offering greater control and flexibility than `curve_fit` for the present application. Although `least_squares` does not directly provide an estimate of the parameter covariance, it returns the Jacobian of the residual function evaluated at the optimal solution. This Jacobian can then be used to estimate parameter uncertainty.

The second strategy consists of implementing a custom optimization routine based on a random-search algorithm. In general terms, this algorithm repeatedly calls the function `int_css`, which performs numerical integration of the differential model at each iteration of the optimization. The function `int_css` invokes the model equations (`css`), integrates them over time, and returns the residuals of the objective function, defined in this case as the differences between the predicted starch

Table P8.1 Relevant parameters and initial conditions

Parameters	Value
k (g/h)	1.34×10^{-4}
k_d (1/h)	2.5×10^{-2}
k_n (g N_I/gsi)	7.0×10^{-4}
$Y_{X/NI}$ (g X/g N_I)	20.4
μ max (1/h)	0.23
Initial conditions	
X (0)	4.7×10^{-3}
U (0)	3.5×10^{-3}
N_I (0)	0
S (0)	0.18
Fermentation time (h)	142

* (gsi): grams of inert support

Table P8.2 Experimental starch measurements (units not defined)

Time (h)	[Starch]	Time (h)	[Starch]
0	0.1131	62	0.0604
2	0.2520	70	0.0411
4	0.2033	78	0.0089
8	0.0918	86	0.0335
12	0.1938	94	0.0274
16	0.1982	102	0.0105
20	0.1402	110	0.0486
25	0.1352	118	0.0174
30	0.1683	126	0.0306
38	0.0760	134	0.0126
46	0.0948	142	0.0203
54	0.0661	–	–

concentrations and the corresponding experimental measurements at the sampling times.

The random-search approach represents a particular case of simulated annealing and belongs to the class of gradient-free optimization methods. As such, it does not rely on analytical or numerical gradients and is less prone to becoming trapped in local optima, making it well suited for problems involving many variables and/or objective functions with multiple local minima.

The Python code developed using the `least_squares function` is presented below:

```
import numpy as np
import matplotlib.pyplot as plt
from scipy.integrate import solve_ivp
from scipy.optimize import least_squares

# Read experimental data (starch = almidon)
almidon_data = np.loadtxt('almidon.txt')
tiempo = almidon_data[:, 0]
almidon = almidon_data[:, 1]

# Model
def css(t, x, params):
    # Parameters
    yxs, ms = params
    mumax, kd, k, yxni, kn = 0.23, 0.025, 1.34e-4, 20.4, 7e-4
    # Differential equations
    # Biomass
    dx1 = mumax * x[2] / (kn + x[2]) * x[0] - kd * x[0]
    # Urea [2] and intermediate nitrogen [3]
    if x[1] > 0:
        dx2 = -k
        dx3 = 0.47 * k - mumax * x[2] / (kn + x[2]) * x[0] / yxni
    else:
        dx2 = 0
        dx3 = 0 - mumax * x[2] / (kn + x[2]) * x[0] / yxni
    # Starch
    dx4 = -mumax * x[2] * x[0] / (kn + x[2]) / yxs - ms * x[0]
    return [dx1, dx2, dx3, dx4]
```

```
# Integration interval for all simulations
t_span = (0, 150)

# Construct residuals: difference between the model and the experimental starch data
def int_css(params):
    X0 = [4.7e-3, 3.5e-3, 0, 0.18]
    # Integrate at the same sampling times as the experimental data
    sol = solve_ivp(css, t_span, X0, args=(params,), t_eval=tiempo)
    return sol.y[3] - almidon

# least_squares optimization routine
initial_guess = [1.0, 0.5]
result = least_squares(int_css, initial_guess)

# Extract estimated parameters
params = result.x

# Comparison between experimental data and model predictions
X0 = [4.7e-3, 3.5e-3, 0, 0.18]
t_sim = np.linspace(0, 150, 100)
almidon_sim = solve_ivp(css, t_span, X0, args=(params,), t_eval=t_sim).y[3]

plt.figure(figsize=(8,6))
plt.plot(tiempo, almidon, 'ko')
plt.plot(t_sim, almidon_sim, 'k-', lw=2)
plt.xlabel('Time (h)', fontsize=14)
plt.ylabel('[Starch]', fontsize=14)
plt.tick_params(axis='both', labelsize=14)
plt.show()

print("Convergence message:", result.message)
print('Estimated parameters:')
print('  yxs:', round(params[0], 2))
print('  ms:', round(params[1], 3))
print(f'- Error (resnorm): {round(2*result.cost, 4)}')
```

After execution of the `least_squares` routine, it is important to examine the output variable `result.message`, which provides information regarding the reason for termination of the algorithm. The reader is referred to the corresponding documentation, which is extensive and highly informative.[4]

Table P8.3 summarizes the main results obtained using both strategies.

Table P8.3 Summary of both optimization strategies

	Least_squares	Simulated–annealing method
Initial values	[1.0 0.5]	[1.0 0.5]
Optimal [$Y_{X/S}$ m_s]	[1.99 0.10]	[1.98 0.10]
Error function (`resnorm`)	0.0243	0.0243

Note that, to obtain the sum of squared errors from `least_squares`, we multiplied the output `result.cost` by two in the implemented code. This is because the algorithm's objective function is defined as one-half of the weighted sum of squared residuals. This design choice is consistent with many optimization libraries, which frequently adopt this one-half factor in the cost function because it simplifies the computation of gradients

[4] https://docs.scipy.org/doc/scipy/reference/generated/scipy.optimize.least_squares.html.

Figure P8.1 shows a simulation of the nutrient concentration compared with the experimental data for the best case obtained using the `least_squares` function.

Figure P8.2 illustrates the evolution of the error function for the simulated–annealing algorithm. As shown, after approximately 300 iterations the algorithm is no longer able to reduce the error. Nevertheless, the final error matches that obtained with `least_squares` to four decimal places. In more complex problems, however, convergence of the error to a stable value does not necessarily guarantee that the global optimum has been attained. In such cases, the parameter vector obtained from the heuristic search may be used as an initial guess for a gradient-based optimization routine, such as `curve_fit` or `least_squares`.

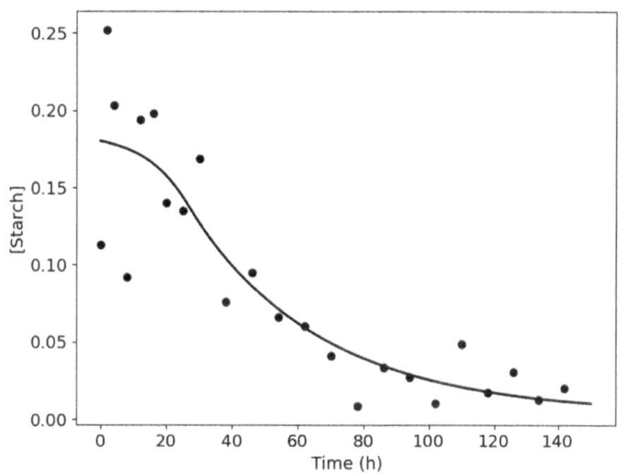

Fig. P8.1 Simulation of starch concentration obtained using the `least_squares` function

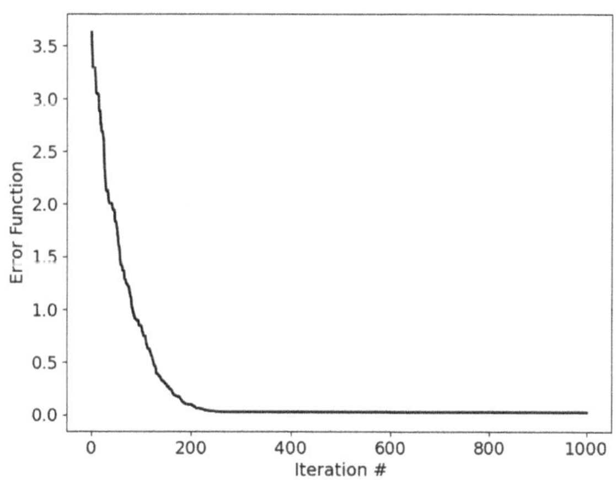

Fig. P8.2 Evolution of the error function for the simulated–annealing algorithm

The Python code for the simulated annealing algorithm is presented below:

```
import numpy as np
import matplotlib.pyplot as plt
from scipy.integrate import solve_ivp
from scipy.optimize import minimize

# Read experimental data (starch = almidon)
almidon_data = np.loadtxt('almidon.txt')  # , delimiter=' ')
tiempo = almidon_data[:, 0]
almidon_exp = almidon_data[:, 1]

# Differential model
def css(t, x, params):
    # Parameters
    yxs, ms = params
    mumax, kd, k, yxni, kn = 0.23, 0.025, 1.34e-4, 20.4, 7e-4
    # Differential equations
    # Biomass
    dx1 = mumax*x[2]/(kn + x[2])*x[0] - kd*x[0]
    # Urea [2] and intermediate nitrogen [3]
    if x[1] > 0:
        dx2 = -k
        dx3 = 0.47*k - mumax*x[2]/(kn + x[2])*x[0]/yxni
    else:
        dx2 = 0
        dx3 = 0 - mumax*x[2]/(kn + x[2])*x[0]/yxni
    # Starch
    dx4 = -mumax*x[2]*x[0]/(kn + x[2])/yxs - ms*x[0]
    return [dx1, dx2, dx3, dx4]

# Integration interval for all simulations
t_span = (0, 150)

# Integrate the differential model at the same experimental times
def int_css(params, tiempo):
    X0 = [4.7e-3, 3.5e-3, 0, 0.18]
    sol = solve_ivp(css, t_span, X0, args=(params,), t_eval=tiempo)
    return sol.y[3]

########################################
# Random optimization (annealing method)
########################################

# First iteration
initial_guess = np.array([1.0, 0.5])
sum_error = np.sum((almidon_exp - int_css(initial_guess, tiempo))**2)

# Algorithm hyperparameters
a, b = -0.025, 0.025
n_iter = 1000
param = initial_guess
ferror = []

# Random-search algorithm
for i in range(n_iter):
    ferror.append(sum_error)
    x = a + (b - a)*np.random.rand(2)
    # Apply the parameters and compute the new error
    p1 = (1 + x)*param
    sum_error_1 = np.sum((almidon_exp - int_css(p1, tiempo))**2)
    # Update the parameters if the error decreases
    if sum_error_1 < sum_error:
        param = p1
        sum_error = sum_error_1

print('Estimated parameters:')
print('  yxs:', round(param[0], 2))
print('  ms:', round(param[1], 3))
```

```
print(f'- Error (resnorm): {round(sum_error, 4)}')

# Plot error function convergence
plt.figure(1)
plt.figure(figsize=(8,6))
plt.plot(range(1, n_iter+1), ferror, 'k', lw=2)
plt.xlabel('Iteration #', fontsize=14)
plt.ylabel('Error Function', fontsize=14)
plt.tick_params(axis='both', labelsize=14)

# Integrate the model using the optimal parameters obtained
X0 = [4.7e-3, 3.5e-3, 0, 0.18]
t_sim = np.linspace(0, 150, 100)
almidon_sim = solve_ivp(css, t_span, X0, args=(param,), t_eval=t_sim).y[3]

# Plot experimental vs simulated data
plt.figure(2)
plt.figure(figsize=(8,6))
plt.plot(tiempo, almidon_exp, 'ko')
plt.plot(t_sim, almidon_sim, 'k-', linewidth=2)
plt.xlabel('Time (h)', fontsize=14)
plt.ylabel('[Starch]', fontsize=14)
plt.tick_params(axis='both', labelsize=14)

plt.show()
```

This problem invites us to reflect on the art of optimization. It is an art because experience shows that having software equipped with optimization routines is not sufficient. Some algorithms perform better than others depending on the structure and complexity of the problem. It is therefore essential to use as much a priori information as possible—for example, selecting initial values with physical or biological meaning (in our case, this involves using typical values of $Y_{X/S}$ and m_s obtained from the specialized literature). Another important aspect is that, in addition to choosing good initial guesses, we should also test multiple starting points. This is because optimization routines are, in general, highly sensitive to the initial point.

More complex situations can arise easily. For example:

- Experimental data are available for both biomass and starch, and the goal is to estimate parameters that appear in both differential equations. In this case, information from both data sets must be incorporated into a single objective (cost) function. To ensure a balanced contribution from each variable, it is advisable to normalize the corresponding errors so that they are comparable in magnitude. For example:

$$\text{Sum of squared errors} = \sum\nolimits_{\text{biomass}} \left(\frac{y_{\text{exp}} - y_{\text{simulated}}}{y_{\text{exp}}} \right)^2 + \sum\nolimits_{\text{starch}} \left(\frac{z_{\text{exp}} - z_{\text{simulated}}}{z_{\text{exp}}} \right)^2.$$

- Additional complexity may arise when the measured variables are sampled at different time intervals. In such cases, the integration routine may be configured to compute the solution at all required sampling times, ensuring that the corresponding simulated values are directly available. Alternatively, the system may be integrated once over the full-time span, and interpolation methods can then be used to estimate the simulated values at specific experimental sampling times.

6.9 Problem 9 Parameter Estimation and Sensitivity in Ordinary Differential Equations

Consider the following homogeneous gas-phase reaction between NO and O_2 (Englezos & Kalogerakis, 2001):

$$2NO + O_2 \leftrightarrow 2NO_2 \tag{P9.1}$$

The kinetics of NO_2 formation can be modeled by the following differential equation:

$$\frac{dx}{dt} = k_1 \cdot (\alpha - x)(\beta - x)^2 - k_2 \cdot x^2, \quad x(0) = 0, \tag{P9.2}$$

where $\alpha = 126.2$, $\beta = 9.19$, and x denotes the concentration of NO_2. Table P9.1 reports the experimentally measured NO_2 concentration as a function of time.

Based on this information, determine the parameters k_1 y k_2, as well as the sensitivity of NO_2 with respect to the estimated parameters as a function of time (dx/dk_1 and $dx/dk_2\ \forall\ t$).

Table P9.1 Experimental data for Eq. (P9.2)

Tiempo	Concentración de NO_2
0	0
1	1.4
2	6.3
3	10.5
4	14.2
5	17.6
6	21.4
7	23.0
9	27.0
11	30.5
14	34.4
19	38.8
24	41.6
29	43.5
39	45.3

Solution

The differential equation has the form:

$$\mathrm{d}x/\mathrm{d}t = f(x; k_1, k_2), \tag{P9.3}$$

where the state variable x is also the measured variable. Since parameter calibration in differential equations was addressed in Problem 8, the present problem focuses on parameter sensitivity, as quantified by the sensitivity matrix *G(t)*. This matrix characterizes the dependence of state variables on the model parameters. For the present problem, *G(t)* is defined as:

$$G(t) = [G_1(t), G_2(t)] = \left[\left(\frac{\partial x}{\partial k_1}\right), \left(\frac{\partial x}{\partial k_2}\right)\right] \tag{P9.4}$$

In general terms, it can be computed as:

$$\frac{\partial}{\partial k}\left(\frac{\mathrm{d}x}{\mathrm{d}t}\right) = \frac{\partial f(x, k)}{\partial k} \tag{P9.5}$$

By rearranging the order of differentiation on the left-hand side of Eq. (P9.5), we obtain:

$$\frac{d}{\mathrm{d}t}\left(\frac{\partial x}{\partial k}\right) = \frac{\partial f}{\partial x} \cdot \frac{\partial x}{\partial k} + \frac{\partial f}{\partial k} \tag{P9.6}$$

Or simply:

$$\frac{\mathrm{d}G(t)}{\mathrm{d}t} = \frac{\partial f}{\partial x} \cdot G(t) + \frac{\partial f}{\partial k}, \tag{P9.7}$$

where *G(t)*, f, x, and k are vectors. The initial condition for $G(t_0)$ is obtained by differentiating the initial condition $x(\mathrm{t_0}) = x_0$ with respect to k. Since the initial state is independent of the parameters, we have:

$$G(t_0) = 0 \tag{P9.8}$$

In the context of our original problem:

$$\frac{\mathrm{d}G_1}{\mathrm{d}t} = \left(\frac{\partial f}{\partial x}\right) \cdot G_1 + \left(\frac{\partial f}{\partial k_1}\right); \quad G_1(0) = 0 \tag{P9.9}$$

$$\frac{\mathrm{d}G_2}{\mathrm{d}t} = \left(\frac{\partial f}{\partial x}\right) \cdot G_2 + \left(\frac{\partial f}{\partial k_2}\right); \quad G_2(0) = 0 \tag{P9.10}$$

Thus, the resulting derivatives are:

$$\left(\frac{\partial f}{\partial x}\right) = -k_1 \cdot (\beta - x)^2 - 2k_1(\alpha - x)(\beta - x) - 2k_2 \cdot x \tag{P9.11}$$

$$\left(\frac{\partial f}{\partial k_1}\right) = (\alpha - x)(\beta - x)^2 \tag{P9.12}$$

$$\left(\frac{\partial f}{\partial k_2}\right) = -x^2. \tag{P9.13}$$

Equations (P9.9) and (P9.10) must be solved simultaneously together with the state Eq. (P9.2). Figure P9.1 illustrates the result of the parameter fitting and the sensitivity of x with respect to k_1 and k_2 obtained by integrating the Eqs. (P9.9) and (P9.10):

It should be emphasized that sensitivity information can also be obtained by replacing analytical derivatives with numerical approximations. This approach has greatly facilitated the development of software tools that automate sensitivity analysis for differential equation models.

The parameters obtained using the `least_squares` optimization routine are:

$$\mathbf{k_1} = 4.6 \cdot 10^{-6}$$
$$\mathbf{k_2} = 2.796 \times 10^{-4}$$

The Python code developed for this case is:

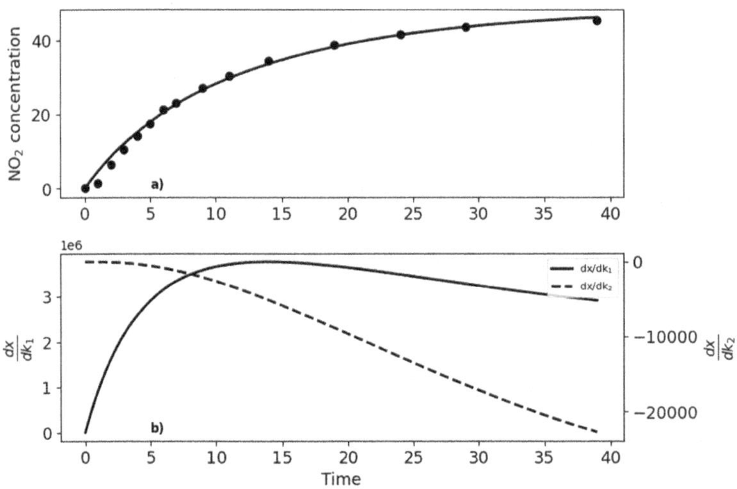

Fig. P9.1 **a** Simulation versus experimental data. **b** Sensitivity of x with respect to k_1 y k_2 as a function of time

```
import numpy as np
import matplotlib.pyplot as plt
from scipy.integrate import solve_ivp
from scipy.optimize import least_squares

# Experimental data
tiempo = np.array([0, 1, 2, 3, 4, 5, 6, 7, 9, 11, 14, 19, 24, 29, 39])
xdata = np.array([0, 1.4, 6.3, 10.5, 14.2, 17.6, 21.4, 23, 27, 30.5, 34.4, 38.8, 41.6,
43.5, 45.3])

# Differential model and dG/dt
def modelo(t, x, params):
    k1, k2 = params
    a = 126.2
    b = 91.9
    dxdt = [k1*(a - x[0])*(b - x[0])**2 - k2*x[0]**2,
            -(k1*(b - x[0])**2 + 2*k1*(a - x[0])*(b - x[0]) + 2*k2*x[0])*x[1] + (a -
x[0])*(b - x[0])**2,
            -(k1*(b - x[0])**2 + 2*k1*(a - x[0])*(b - x[0]) + 2*k2*x[0])*x[2] -
x[0]**2]
    return dxdt

# Model integration function
def integra(params):
    # Initial conditions
    X0 = [0, 0, 0]
    # Integration at the experimental sampling times
    sol = solve_ivp(modelo, (min(tiempo), max(tiempo)), X0, args=(params,),
t_eval=tiempo, method='DOP853')
    return sol.y[0] - xdata

# Least-squares optimization routine
initial_guess = [1e-5, 1e-3]
results = least_squares(integra, initial_guess)

# Output of the results
print("Message of convergence:", results.message)
print(f'- Estimated parameters (k1, k2): {np.round(results.x, 7)}')

# Model simulation using the estimated parameters
t_sim = np.linspace(0, max(tiempo), 100)
solution = solve_ivp(modelo, (min(tiempo), max(tiempo)), [0, 0, 0], args=(results.x,),
t_eval=t_sim)

# Plotting the results
plt.figure(figsize=(8, 6))
plt.subplot(2, 1, 1)
plt.plot(tiempo, xdata, 'ko')
plt.plot(t_sim, solution.y[0], 'k-', linewidth=2)
plt.ylabel(r'NO$_2$ concentration', fontsize=13)
plt.text(5, 0, "a)", fontweight='bold')
plt.tick_params(axis='both', labelsize=13)

plt.subplots_adjust(hspace=0.3)

plt.subplot(2, 1, 2)
ax1 = plt.gca()
ax2 = ax1.twinx()
ax1.plot(t_sim, solution.y[1], 'k', linewidth=2, label=r'dx/dk$_1$')
ax2.plot(t_sim, solution.y[2], 'k--', linewidth=2, label=r'dx/dk$_2$')
ax1.set_xlabel('Time', fontsize=13)
ax1.text(5, 0, "b)", fontweight='bold')
ax1.set_ylabel(r'$\frac{dx}{dk_1}$', color='k', fontsize=16)
ax2.set_ylabel(r'$\frac{dx}{dk_2}$', color='k', fontsize=16)
ax1.tick_params(axis='both', labelsize=13)
ax2.tick_params(axis='both', labelsize=13)

lines, labels = ax1.get_legend_handles_labels()
lines2, labels2 = ax2.get_legend_handles_labels()
ax2.legend(lines + lines2, labels + labels2, loc='best', fontsize=8)

plt.show()
```

Understanding the sensitivity of a system with respect to its parameters is highly valuable for parameter analysis, as it enables, among other applications, the computation of the Fisher information matrix. This matrix allows us to determine whether a system of differential equations is locally identifiable (for example, whether the estimated parameter set is unique), given the model structure, the estimated parameters, and the experimental data used for calibration.

To extend this approach to systems with additional differential equations and parameters, it is convenient to adopt a matrix formulation. Consider, in general terms, a system consisting of two differential equations with state variables x_I and x_2, and two parameters k_1 and k_2:

$$\begin{aligned} \frac{dx_1}{dt} &= f_1(x_1, x_2; k_1, k_2) \\ \frac{dx_2}{dt} &= f_2(x_1, x_2; k_1, k_2) \end{aligned}. \tag{P9.14}$$

The sensitivity matrix, $G(t)$, is a 2×2 matrix:

$$G(t) = [g_1(t), g_2(t)] = \left[\left(\frac{\partial x}{\partial k_1}\right), \left(\frac{\partial x}{\partial k_2}\right)\right] \tag{P9.15}$$

$$\begin{bmatrix} G_{11}(t) & G_{12}(t) \\ G_{21}(t) & G_{22}(t) \end{bmatrix} = \begin{bmatrix} \frac{\partial x_1}{\partial k_1} & \frac{\partial x_1}{\partial k_2} \\ \frac{\partial x_2}{\partial k_1} & \frac{\partial x_2}{\partial k_2} \end{bmatrix}, \tag{P9.16}$$

where G_{ij} (with i denoting the state variable and j the parameter). The corresponding sensitivity differential equations can then be written as:

$$\begin{aligned} \frac{dg_1(t)}{dt} &= \frac{\partial f}{\partial t} \cdot g_1(t) + \left(\frac{\partial f}{\partial k_1}\right); \quad g_1(t) = 0 \\ \frac{dg_2(t)}{dt} &= \frac{\partial f}{\partial t} \cdot g_2(t) + \left(\frac{\partial f}{\partial k_2}\right); \quad g_2(t) = 0 \end{aligned}. \tag{P9.17}$$

In matrix form:

$$\begin{bmatrix} \frac{dG_{11}}{dt} \\ \frac{dG_{21}}{dt} \end{bmatrix} = \begin{bmatrix} \frac{\partial f_1}{\partial x_1} & \frac{\partial f_1}{\partial x_2} \\ \frac{\partial f_2}{\partial x_1} & \frac{\partial f_2}{\partial x_2} \end{bmatrix} \cdot \begin{bmatrix} G_{11} \\ G_{21} \end{bmatrix} + \begin{bmatrix} \frac{\partial f_1}{\partial k_1} \\ \frac{\partial f_2}{\partial k_1} \end{bmatrix}; \quad G_{11}(t_0) = G_{21}(t_0) = 0 \tag{P9.18}$$

$$\begin{bmatrix} \frac{dG_{12}}{dt} \\ \frac{dG_{22}}{dt} \end{bmatrix} = \begin{bmatrix} \frac{\partial f_1}{\partial x_1} & \frac{\partial f_1}{\partial x_2} \\ \frac{\partial f_2}{\partial x_1} & \frac{\partial f_2}{\partial x_2} \end{bmatrix} \cdot \begin{bmatrix} G_{12} \\ G_{22} \end{bmatrix} + \begin{bmatrix} \frac{\partial f_1}{\partial k_2} \\ \frac{\partial f_2}{\partial k_2} \end{bmatrix}; \quad G_{21}(t_0) = G_{22}(t_0) = 0. \tag{P9.19}$$

Finally, the differential equations that must be added to the original system (Eq. (P9.14)) to integrate them simultaneously are:

$$
\begin{aligned}
\frac{\mathrm{d}G_{11}}{\mathrm{d}t} &= \left(\frac{\partial f_1}{\partial x_1}\right)G_{11} + \left(\frac{\partial f_1}{\partial x_2}\right)G_{21} + \frac{\partial f_1}{\partial k_1}; \quad G_{11}(0) = 0 \\
\frac{\mathrm{d}G_{21}}{\mathrm{d}t} &= \left(\frac{\partial f_2}{\partial x_1}\right)G_{11} + \left(\frac{\partial f_2}{\partial x_2}\right)G_{21} + \frac{\partial f_2}{\partial k_1}; \quad G_{21}(0) = 0 \\
\frac{\mathrm{d}G_{12}}{\mathrm{d}t} &= \left(\frac{\partial f_1}{\partial x_1}\right)G_{12} + \left(\frac{\partial f_1}{\partial x_2}\right)G_{22} + \frac{\partial f_1}{\partial k_2}; \quad G_{12}(0) = 0 \\
\frac{\mathrm{d}G_{22}}{\mathrm{d}t} &= \left(\frac{\partial f_2}{\partial x_1}\right)G_{12} + \left(\frac{\partial f_2}{\partial x_2}\right)G_{22} + \frac{\partial f_2}{\partial k_2}; \quad G_{22}(0) = 0
\end{aligned} \tag{P9.20}
$$

In a more general form, the differential equation governing a generic element G_{ij} of the sensitivity matrix, now of dimension [n equations] $\times$ [parameters], can be written as:

$$
\frac{\mathrm{d}G_{ij}}{\mathrm{d}t} \equiv \frac{d}{\mathrm{d}t}\left(\frac{\partial x_i}{\partial k_j}\right) = \sum_{k=1}^{n}\left(\frac{\partial f_i}{\partial x_k}\right)\left(\frac{\partial x_k}{\partial k_j}\right) + \frac{\partial f_i}{\partial k_j} \equiv \sum_{k=1}^{n}\left(\frac{\partial f_i}{\partial x_k}\right)G_{kj} + \frac{\partial f_i}{\partial k_j}. \tag{P9.21}
$$

A practical application of these concepts is provided by the HIPPO tool, which is used for calibration of complex bioreactor models (Sánchez et al., 2014). HIPPO employs an iterative heuristic optimization procedure to estimate model parameters and identify those exhibiting acceptable behavior. In this context, acceptable behavior refers to parameters that display sufficient sensitivity, exert a meaningful influence on the model predictions, and possess statistical significance.

In parameter estimation problems of this type, it is essential to avoid parameters with low sensitivity—those that have little effect on the model response—or low statistical significance, which is typically associated with high variability in their estimates. Such issues are common in complex models and necessitate the use of specialized tools to assess the reliability and validity of the resulting solutions.

6.10 Problem 10 Heat Transfer in a Circular Fin: Boundary Value Problem

The circular fin shown in Fig. P10.1 is welded to a pipe whose wall temperature is known and equal to T_0. This fin can be modeled by the following second-order ordinary differential equation (Bird et al., 2007):

$$
\frac{\mathrm{d}^2T}{\mathrm{d}r^2} + \frac{1}{r}\cdot\frac{\mathrm{d}T}{\mathrm{d}r} - \frac{2\cdot h}{B\cdot k}\cdot(T - T_a) = 0 \tag{P10.1}
$$

where the boundary conditions for the physical problem are given by:

$r = R_0$:

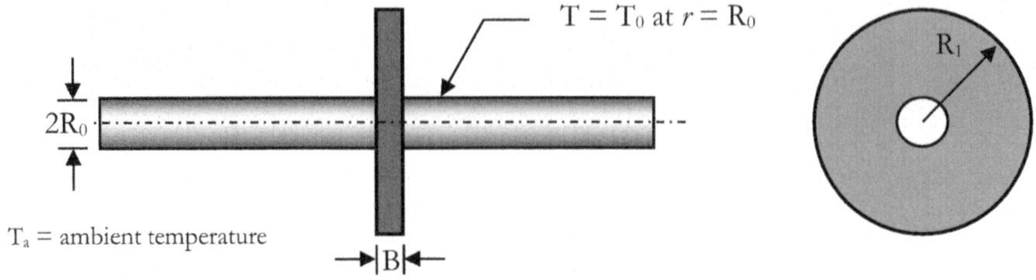

Fig. P10.1 Side and front views of the circular fin and the pipe

$$T(r = R_0) = T_0 \tag{P10.2}$$

$r = R_1$:

$$-k \cdot \left.\frac{\mathrm{d}T}{\mathrm{d}r}\right|_{R_1} = h \cdot (T(r = R_1) - T_a) \tag{P10.3}$$

Data: $B = 1$ mm; $h = 10$ W/m^2 K; $R_0 = 1$ cm; $R_1 = 10$ cm; $T_a = 10$ °C; $T_0 = 100$ °C.

For the conditions described above, determine whether differences exist in the temperature profiles over the radial domain $R_0 \leq r \leq R_1$ for fins fabricated from the materials listed in Table P10.1.

To address this problem:

(a) For each material, plot the temperature as a function of the radial coordinate r. Discuss the effect of thermal conductivity on the temperature profile of the fin.
(b) For each material, plot the cumulative heat loss (W) through both faces as a function of the radial position r.

Solution

The first difficulty encountered is the order of the differential equation (Eq. (P10.1)). Up to this point, only first-order differential equations have been considered. To employ the integrators available in SciPy, a change of variables must be introduced to convert an nth order differential equation into a system of n first-order equations. In the present case, this transformation introduces the auxiliary variables z_1 and z_2:

Table P10.1 Thermal conductivities of selected metals

Material	Thermal conductivity (W/m K)
Copper	382
Aluminum	204
Iron	62.3
Steel	14.9

$$z_1 = T \tag{P10.4}$$

$$z_2 = \mathrm{d}T/\mathrm{d}r. \tag{P10.5}$$

If the governing equation were of higher order, additional auxiliary variables would be introduced in an analogous manner (e.g., $z_3 = \mathrm{d}^2y/\mathrm{d}x^2$, $z_4 = \cdots$).

Differentiating Eqs. (P10.4) and (P10.5), the expression for $\mathrm{d}T/\mathrm{d}r$ is substituted into Eq. (P10.4), while the term $\mathrm{d}z/\mathrm{d}r = \mathrm{d}^2T/\mathrm{d}r^2$, obtained by isolating the second derivative from Eq. (P10.1), is substituted into Eq. (P10.5). In this way, the original higher-order differential equation is recast as a system of first-order equations in the new variables:

$$\frac{\mathrm{d}z_1}{\mathrm{d}r} = z_2 \tag{P10.6}$$

$$\frac{\mathrm{d}z_2}{\mathrm{d}r} = -\frac{1}{r} \cdot z_2 + \frac{2h}{B \cdot k} \cdot (z_1 - T_a). \tag{P10.7}$$

As a result, the system of Eqs. (P10.6) and (P10.7) can now be integrated using the `solve_ivp` routine from the SciPy library.

The second difficulty in this problem arises from the boundary condition at $T(r = R_1)$. To integrate Eqs. (P10.6) and (P10.7), we must provide two boundary conditions: one for the temperature and one for its radial derivative, both at the same spatial location of the fin. The natural choice is $r = R_0$, since $T(r = R_0)$ is known. However, the second boundary condition, $\mathrm{d}T/\mathrm{d}r$ evaluated at $r = R_0$, is not known a priori.

Formally, problems of this type are classified as boundary value problems. One approach for determining the missing boundary condition is to employ a dedicated boundary value solver, such as the `solve_bvp` routine from the SciPy library. An alternative approach is to formulate an equivalent optimization problem in which the following expression is minimized:

$$\left(k \cdot \left.\frac{\mathrm{d}T}{\mathrm{d}r}\right|_{R_1} + h \cdot (T(r = R_1) - T_a)\right)^2 = 0. \tag{P10.8}$$

Expression (P10.8) is obtained from the boundary condition of the physical problem (Eq. (P10.3)) evaluated at $r = R_1$, rewritten such that it is equal to zero. In essence, the objective is to determine a boundary condition at $r = R_0$ that ensures satisfaction of the boundary condition at $r = R_1$. Squaring Eq. (P10.8) ensures that the minimization procedure searches for a minimum value of zero, rather than diverging toward negative infinity.

To solve all cases (all metals) within a single simulation, we included a summation over all metals. Of course, each metal could have been treated independently, but our goal here is to demonstrate that a combined approach is also feasible, while also saving space in the presentation.

To solve this problem, the `minimize`[5] routine from the SciPy library is employed. This routine is a versatile optimization tool that offers a wide range of algorithms and can be applied to problems ranging from single-variable scalar minimization to complex multivariable optimization.

In the present case, `minimize` repeatedly calls the numerical integrator, which in turn evaluates the objective (cost) function to be minimized, defined as the sum of squared residuals (Eq. (P10.9)) for each metal. The optimization procedure requires initial guesses for the temperature derivatives at $r = R_0$, which may be selected by trial and error.

The optimization results are summarized in Table P10.2.

Figure P10.2 shows the temperature profile obtained for each metal.

In Fig. P10.2, we observe that as the thermal conductivity of the metal decreases, the temperature profile drops more rapidly. For example, steel—the poorest conductor in Table P10.1—reaches the lowest temperature at $r = R_1$. At the opposite end, copper—the best conductor listed in Table P10.2— reaches a final value of 70 °C. This behavior is consistent with the definition of thermal conductivity, which represents the ability of a material to transport heat.

In terms of which material dissipates the greatest amount of heat, it is sufficient to analyze Newton's law of cooling:

Table P10.2 Summary of the optimization results

	Temperature derivative value
Copper	– 1669.0
Aluminum	– 2506.5
Iron	– 4325.4
Steel	– 6694.1

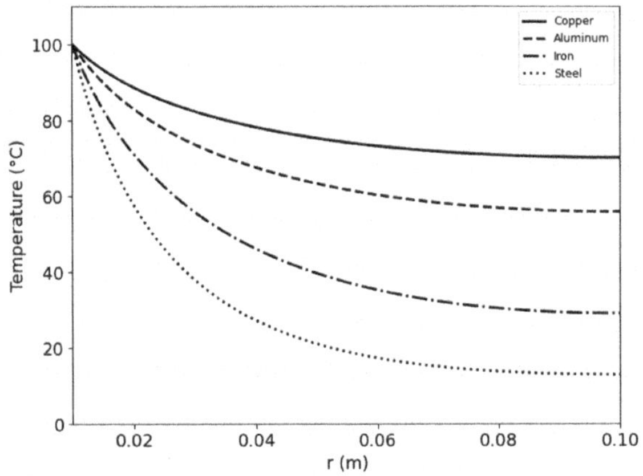

Fig. P10.2 Temperature profiles for the four metals as a function of the radius r

[5] https://docs.scipy.org/doc/scipy/reference/generated/scipy.optimize.minimize.html.

$$q|_r = h \cdot (T(r) - T_a). \tag{P10.9}$$

The heat loss is proportional to the temperature difference between the fin and the surrounding air. For the conditions considered, copper is the material that dissipates heat most effectively. To compute the total heat loss, an additional differential equation may be appended to the existing system of equations:

$$\frac{dz_3}{dr} = 4\pi \cdot r \cdot h \cdot (z_1 - T_a). \tag{P10.10}$$

This expression accounts for both the upper and lower surfaces, since the fin loses heat through both. Figure P10.3 illustrates the accumulated heat loss between R_0 and r for the different metals. The results provide a graphical illustration of the behavior predicted by Newton's law of cooling.

The Python code used to solve this problem is provided below:

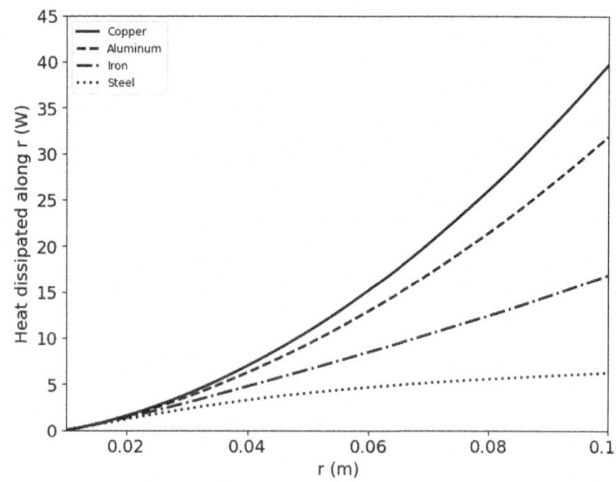

Fig. P10.3 Heat loss from the fin as a function of r

```
import numpy as np
import matplotlib.pyplot as plt
from scipy.integrate import solve_ivp
from scipy.optimize import minimize

# Parameters for both functions
T0 = 373.15        # K
h = 10             # W/m^2·K
Ta = 273.15 + 10   # K
k = [382, 204, 62.3, 14.9]  # [0=copper, 1=aluminum, 2=iron, 3=steel] W/m·K

# Fin model with change of variables
def model(r, z, k):
    B = 0.001  # m
    dz1 = z[1]
    dz2 = -1/r * z[1] + 2 * h / (B * k) * (z[0] - Ta)
    dz3 = 4 * np.pi * r * h * (z[0] - Ta)
    return [dz1, dz2, dz3]

# Model that integrates the fin equation and returns fcosto, which represents
# the boundary condition at r = R1 for the four materials: conduction equals convection
def fin(param, k):
    r = (0.01, 0.1)
    rvalues = np.linspace(0.01, 0.1, 100)
    T1 = solve_ivp(model, r, [T0, param[0], 0], args=(k[0],), t_eval=rvalues).y
    T2 = solve_ivp(model, r, [T0, param[1], 0], args=(k[1],), t_eval=rvalues).y
    T3 = solve_ivp(model, r, [T0, param[2], 0], args=(k[2],), t_eval=rvalues).y
    T4 = solve_ivp(model, r, [T0, param[3], 0], args=(k[3],), t_eval=rvalues).y

    fcosto = 1000 * (
        (k[0] * T1[1, -1] + h * (T1[0, -1] - Ta))**2 +
        (k[1] * T2[1, -1] + h * (T2[0, -1] - Ta))**2 +
        (k[2] * T3[1, -1] + h * (T3[0, -1] - Ta))**2 +
        (k[3] * T4[1, -1] + h * (T4[0, -1] - Ta))**2
    )
    return fcosto

# Optimization routine
# To satisfy the condition conduction = convection at the fin tip, we seek the optimal
vector "param"
initial_guess = [-1000] * 4
result = minimize(fin, initial_guess, args=(k,), method='Powell', tol=1e-9)
print(result)
params_opt = result.x
print(f'\nOptimized initial conditions: {np.round(params_opt, 2)}')
print(f'Cost function evaluated at the optimum: {np.round(result.fun, 8)}')

# Plotting the temperature and total heat loss for the four cases
# Line style configuration
line_styles = ['-', '--', '-.', ':']

# Integration range
r = (0.01, 0.1)
rvalues = np.linspace(0.01, 0.1, 100)

materials = ['Copper', 'Aluminum', 'Iron', 'Steel']
for i in range(4):
    T = solve_ivp(model, r, [T0, params_opt[i], 0], args=(k[i],), t_eval=rvalues).y

    # Temperature
    plt.figure(1, figsize=(8, 6))
    plt.plot(rvalues, T[0] - 273.15, color='k',
             label=materials[i], linestyle=line_styles[i], lw=2)
    plt.xlabel('r (m)', fontsize=14)
    plt.xlim(0.01, 0.1)
    plt.ylabel('Temperature (°C)', fontsize=14)
    plt.ylim(0, 110)
    plt.legend(loc='best', fontsize=9)
    plt.tick_params(axis='both', labelsize=14)
```

```
    # Heat dissipated
    plt.figure(2, figsize=(8, 6))
    plt.plot(rvalues, T[2], color='k',
             label=materials[i], linestyle=line_styles[i], lw=2)
    plt.xlabel('r (m)', fontsize=14)
    plt.xlim(0.01, 0.1)
    plt.ylabel('Heat dissipated along r (W)', fontsize=14)
    plt.ylim(0, 45)
    plt.legend(loc='best', fontsize=9)
    plt.tick_params(axis='both', labelsize=14)

plt.show()
```

The reader is encouraged to solve this problem using the `solve_bvp` integrator from the SciPy library. This function numerically solves a system of first-order ODEs subject to two-point boundary conditions.

6.11 Problem 11 Rotating Cylinder Between Two Fluids

A hollow cylinder, closed at both ends, is partially immersed between two fluids (problem adapted from (Luyben, 1973). The first half is submerged in a cooling liquid, while the other half is exposed to a stream of hot air, as illustrated in the figure. The cylinder rotates at an angular velocity ω, and its dimensions are length L, wall thickness d, and outer radius R. The heat transfer coefficients in the cooling and heating regions are constant and equal to h_c and h, respectively. It may be assumed that the cylinder operates under steady-state conditions and that the fluid temperatures T_h and T_c remain constant along the cylinder surface. In practice, the system reaches this regime only after a thermal startup period; if the wall thermal response time is comparable to (or longer than) the operating time, the cylinder may remain in transient operation, and the steady-state model may be invalid. In addition, assume that the cylinder has end caps and that no heat is gained or lost through its interior. Based on this information, plot the temperature profiles between 0 and 360°.

Data: $cp = 100\,\text{J/kg K}$; $d = 0.01$ m; h (hot air) $= 4\,\text{W/m}^2\text{K}$; h_c (coolant) $= 15\,\text{W/m}^2 \times \text{K}$; $k = 7\,\text{W/m} \times \text{K}$; $L = 1.5\,\text{m}$; $R = 0.5\,\text{m}$; $T_c = 318.16\,\text{K}$; $T_h = 356.16\,\text{K}$; $\rho = 2500\,\text{kg/m}^3$; $\omega = 0.005$ rad/s (Fig. P11.1).

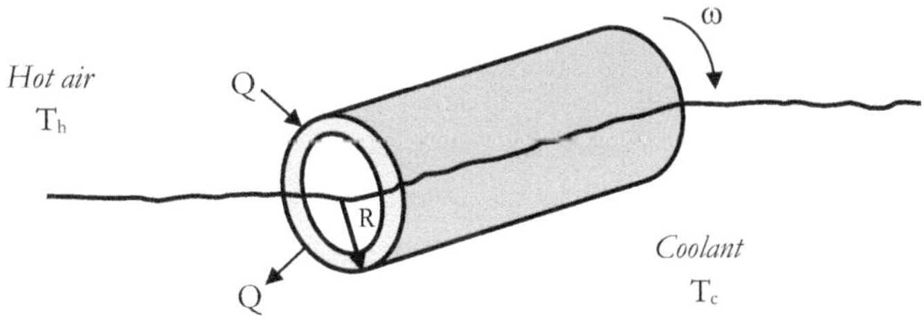

Fig. P11.1 Cylinder rotating between the two fluids

Solution

Given the drum radius ($R = 50$ cm) and wall thickness ($d = 1$ cm), it is reasonable to neglect temperature variations in the radial direction. With this assumption, a differential energy balance may be formulated using fixed spatial coordinates (Fig. P11.2)

The differential steady-state energy balance is:

$$(d \cdot L) \cdot \left(q_c|_{\theta+d\theta} - q_c|_{\theta} \right) + (R \cdot d\theta \cdot L) \cdot q_h|_{\theta+\theta/2} + \dot{m} \cdot cp \cdot [(T(\theta) - T_{ref}) - (T(\theta + d\theta) - T_{ref})] = 0 \quad \text{(P11.1)}$$

where

q_c heat transfer by conduction per unit area

q_h heat transfer by convection per unit area

$\dot{m}$ mass flow rate of the drum material crossing the control-volume boundary per unit time

T_{ref} reference temperature.

In this case, the mass flow rate $\dot{m}$ is given by:
$\dot{m}$ = (velocity) (cross-sectional area) (material density)

$$\dot{m} \approx (\omega \cdot R) \cdot (d \cdot L) \cdot \rho \quad \text{(P11.2)}$$

Substituting Eq. (P11.2) into Eq. (P11.1) and dividing the resulting expression by $Rd\theta$:

$$(d \cdot L) \cdot \frac{q_c|_{\theta+d\theta} - q_c|_{\theta}}{Rd\theta} + L \cdot q_h|_{\theta+\theta/2} + (\omega \cdot R \cdot d \cdot L \cdot \rho) \cdot cp \frac{T(\theta) - T(\theta + d\theta)}{Rd\theta} = 0. \quad \text{(P11.3)}$$

Now, taking the limit as $d\theta \to 0$:

$$\left(\frac{d \cdot L}{R} \right) \cdot \frac{dq_c}{d\theta} + L \cdot q_h - (\omega \cdot d \cdot L \cdot \rho) \cdot cp \cdot \frac{dT}{d\theta} = 0 \quad \text{(P11.4)}$$

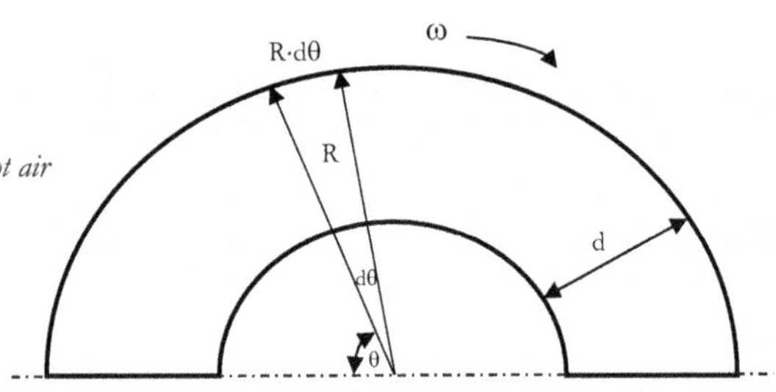

Fig. P11.2 Energy balance for the differential section of the drumexposed to the hot air (the wall thickness is not shown to scale)

The conductive heat-transfer term (q_c) is given by Fourier's law, and the convective heat-transfer term (q_h) is modeled using Newton's law of cooling. Substituting these expressions into Eq. (P11.4) and performing some algebraic rearrangement, we obtain:

$$\frac{1}{R^2}\frac{d^2T}{d\theta^2} + \frac{\omega \cdot \rho \cdot cp}{k} \cdot \frac{dT}{d\theta} - \frac{h}{d \cdot k} \cdot (T_h - T) = 0. \tag{P11.5}$$

For the section of the drum submerged in the coolant, it is straightforward to show that the resulting equation is:

$$\frac{1}{R^2}\frac{d^2T}{d\theta^2} - \frac{\omega \cdot \rho \cdot cp}{k} \cdot \frac{dT}{d\theta} - \frac{h_c}{d \cdot k} \cdot (T - T_c) = 0. \tag{P11.6}$$

Equations (P11.5) and (P11.6) account for conduction, convection, and the rotational transport of the cylinder. To assess the relative importance of thermal conduction (second derivative in Eqs. (P11.5) and (P11.6)) compared to the convective transport term (associated with drum rotation), we introduce the dimensionless Peclet number (Pe), defined as:

$$\text{Pe} = \frac{\text{velocity } x \text{ characteristic length}}{\text{thermal difusivity}} = \frac{(\omega R) \cdot (\pi R)}{\alpha}. \tag{P11.7}$$

Evaluation of the problem data indicates that the Peclet number is much greater than unity ($\text{Pe} \approx 140$), implying that convective transport dominates over conductive transport. Consequently, the thermal conduction term in the circumferential (rotational) direction of the cylinder may be neglected. Under this assumption, the governing equations for both sections reduce to:

$$\frac{dT}{d\theta} = \frac{h \cdot (T_h - T)}{cp \cdot d \cdot \rho \cdot \omega} \tag{P11.8}$$

$$\frac{dT}{d\theta} = \frac{-h_c \cdot (T - T_c)}{cp \cdot d \cdot \rho \cdot \omega}. \tag{P11.9}$$

Both differential equations can be solved analytically, and this exercise is left to the reader. Instead, we will use a slightly longer—yet more instructive—approach from a programming perspective.

Each of the differential equations (Eqs. (P11.8) and (P11.9)) requires a boundary condition specifying the temperature at a particular angular position on the cylinder. For the drum surface exposed to hot air, it is convenient to prescribe $T(\theta_1 = 0)$; similarly, for the section in contact with the coolant, it is convenient to specify $T(\theta_2 = 0)$. The locations at which these temperatures are defined are shown in Fig. P11.3:

Sections (1) and (2) of Fig. P11.3 are connected, which allows us to establish the following continuity condition at point (1):

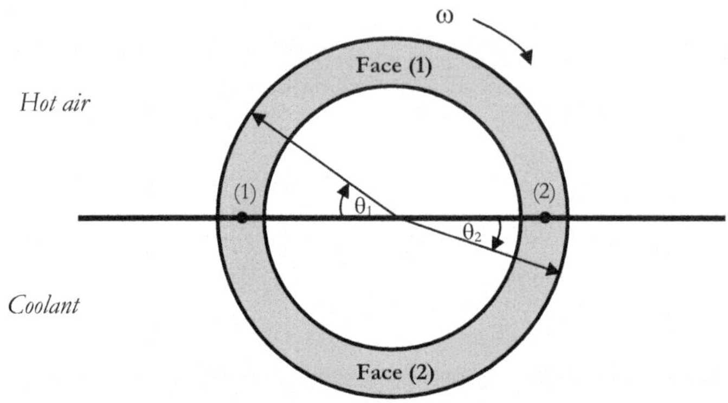

Fig. P11.3 Boundary conditions for both sections of the drum

$$T(\theta_1 = 0) = T(\theta_2 = \pi). \qquad \text{(P11.10)}$$

To obtain the temperature profile, the strategy is to define a function to be minimized that incorporates both previously discussed conditions. The resulting function is as follows:

$$\text{Min} \rightarrow (T(\theta_1 = 0) - T(\theta_2 = \pi))^2 \qquad \text{(P11.11)}$$

Note that Eq. (P11.11) depends only on the unknown boundary condition $T(\theta_1 = 0)$. Once this temperature is specified, the model for Sect. 1 can be integrated. The resulting temperature $T(\theta_1 = \pi)$ can then be used as the boundary condition for the second differential equation. After integration of the second model, the temperature at $\theta_2 = \pi$ must match the value initially specified for $T(\theta_1 = 0)$.

The code developed to solve this problem is organized into three components. The main function executes the optimization routine. The second part is the function `int_tambor`, which integrates the differential equations corresponding to `face1` and `face2` and returns the value of the objective function (Eq. (P11.11)). The third component consists of the differential equations (P11.9) and (P11.10), together with the respective differential equations describing the gain or loss of energy (per unit area) for the exposed section.

In summary, the optimization routine repeatedly calls the integrator for Eqs. (P11.8) and (P11.9) and evaluates the objective function. If the resulting value does not satisfy the convergence criterion, the integrator is invoked again with an updated parameter value. This process is repeated automatically until the convergence criteria imposed by the `minimize` routine are satisfied.

The Python code used in this case is:

```
import numpy as np
import matplotlib.pyplot as plt
from scipy.optimize import minimize
from scipy.integrate import solve_ivp

# Parameters
ro = 2500; cp = 100; hc = 15
h = 4; d = 0.01; w = 0.005
Tc = 318.16; Th = 356.16

# Model for face 1
def face1(ang, z):
    # Differential equation: face exposed to hot air
    dz1 = h * (Th - z) / (cp * d * ro * w)
    return dz1

# Model for face 2
def face2(ang, z):
    # Differential equation: face exposed to the coolant
    dz2 = -hc * (z - Tc) / (cp * d * ro * w)
    return dz2

# Integration times for face 1
t1 = (0, np.pi)
t1values = np.linspace(0, np.pi, 100)

# Integration times for face 2
t2 = (np.pi, 2 * np.pi)
t2values = np.linspace(np.pi, 2 * np.pi, 100)

# Cost function to be minimized
def int_tambor(condInicial):
    T1 = solve_ivp(face1, t1, condInicial, t_eval=t1values).y
    T2 = solve_ivp(face2, t2, y0=[T1[0, -1]], t_eval=t2values).y
    # Cost function
    return (condInicial - T2[0, -1])**2

# Optimization routine
result = minimize(int_tambor, x0=325)
print(result)

# Simulation to generate data for plotting
condInicial_optima = result.x
T1 = solve_ivp(face1, t1, condInicial_optima, t_eval=t1values).y
T2 = solve_ivp(face2, t2, y0=[T1[0, -1]], t_eval=t2values).y

# Plotting
plt.figure(figsize=(8, 6))
plt.plot(t1values / np.pi * 180, T1[0], 'k')
plt.plot(t2values / np.pi * 180, T2[0], 'k')
plt.xlabel('Angle', fontsize=14)
plt.ylabel('Temperature (K)', fontsize=14)
plt.tick_params(axis='both', labelsize=14)

plt.show()
```

The solution obtained for the boundary condition is:

$$T(\theta_1 = 0) = 318.7\ \mathrm{K}$$

An identical value is obtained when solving the differential equations analytically (this short exercise is left to the reader):

$$T(\theta_1 = 0) = \frac{T_c - T_h \cdot e^{\left(\frac{-\pi \cdot (h + h_C)}{cp \cdot d \cdot \omega \cdot \rho}\right)} + (T_h - T_c) \cdot e^{\left(\frac{-\pi \cdot h_C}{cp \cdot d \cdot \omega \cdot \rho}\right)}}{1 - e^{\left(\frac{-\pi \cdot (h + h_C)}{cp \cdot d \cdot \omega \cdot \rho}\right)}} \tag{P11.12}$$

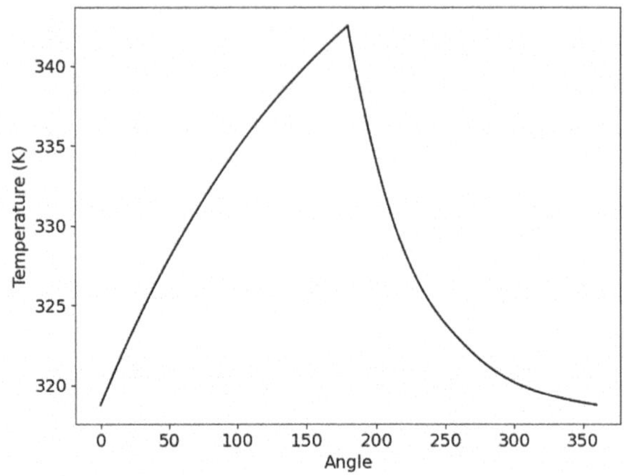

Fig. P11.4 Temperature profile obtained using the calculated boundary condition

$$T(\theta_1 = 0) = 318.7 \text{ K}.$$

The resulting temperature profile, obtained by evaluating the computed boundary-condition value, is (Fig. P11.4).

Finally, the reader is encouraged to re-solve this problem including the heat-conduction term and to compare the resulting solution with that obtained here, in which this term was neglected.

6.12 Problem 12 Application of Finite Differences to Partial Differential Equations (PDEs)

A thick copper plate is initially at a uniform temperature of 20 °C (Incropera et al., 2007). Suddenly, its surface is exposed to a thermal radiation source (q_0''), such that the net heat flux is maintained constant at 3×10^5 W/m^2. Determine the temperature at the irradiated surface and at an interior point located 150 mm from the surface after 120 s have elapsed (Fig. P12.1).

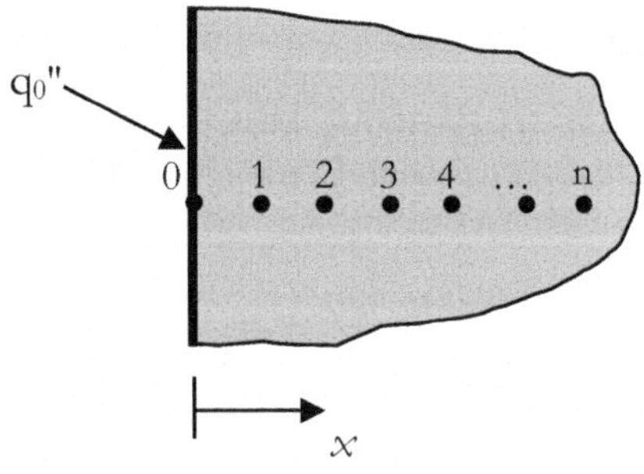

Fig. P12.1 Schematic of the copper plate and the first nodes

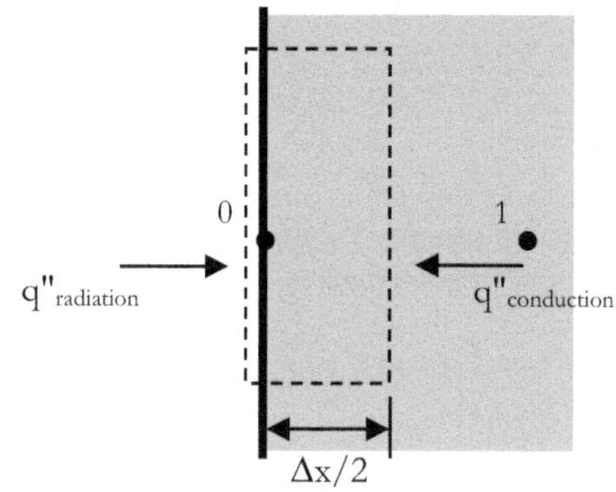

Fig. P12.2 Energy balance at node 0

Data: Thermal conductivity $k = 401\ \mathrm{W/m\,K}$; thermal diffusivity[6] $\alpha = 1.17 \times 10^{-4}\ \mathrm{m^2/s}$.

Solution

Figure P12.2 shows the control volume around node 0, where the first energy balance is applied:

This balance expresses that the net heat input by radiation, minus the heat loss by conduction, is equal to the rate of energy accumulation within the control volume. Mathematically, this relationship can be written as:

$$q''_0 + k \cdot \frac{T_1^{p+1} - T_0^{p+1}}{\Delta x} = \rho \cdot \frac{\Delta x}{2} \cdot cp \cdot \frac{T_0^{p+1} - T_0^{p}}{\Delta t} \tag{P12.1}$$

Taking $\Delta t = 24$ seg and $\Delta x = 75$ mm (chosen for illustrative purposes), Eq. (P12.1) reduces to:

$$2 \cdot T_0^{p+1} - T_1^{p+1} = 56.1 + T_0^{p} \tag{P12.2}$$

For the interior nodes (i.e., 1, 2, …), the implicit finite-difference form of Fourier's equation is employed. In this formulation, the time derivative is evaluated between the current step p and the future step $p + 1$, while the indices m and n denote the spatial position within the grid. Accordingly, the discretized equation is given by:

$$\frac{1}{\alpha} \cdot \frac{T_{m,n}^{p+1} - T_{m,n}^{p}}{\Delta t} = \frac{T_{m+1,n}^{p+1} + T_{m-1,n}^{p+1} - 2 \cdot T_{m,n}^{p+1}}{(\Delta x)^2} + \frac{T_{m,n+1}^{p+1} + T_{m,n-1}^{p+1} - 2 \cdot T_{m,n}^{p+1}}{(\Delta y)^2}. \tag{P12.3}$$

Given the symmetry of the problem, for an interior node we apply Eq. (P12.3) only in the x-direction of the plane (i.e., n does not change). Under this assumption,

[6] Thermal diffusivity is defined as k/ρ cp.

the resulting expression is:

$$-T_{m-1}^{p+1} + 4 \cdot T_m^{p+1} - T_{m+1}^{p+1} = 2 \cdot T_m^p \tag{P12.4}$$

The resulting nine nodes can be assembled into the following coefficient matrix. Each row contains the coefficients of the energy-balance equations evaluated at the unknown time step $p + 1$. Thus, the first row (node 0) includes coefficients 2 and − 1, which correspond to the temperatures on the left-hand side of Eq. (P12.2). Similarly, the remaining rows (nodes 1, 2, ..., 8) include only the coefficients from Eq. (P12.4) for step $p + 1$.

$$\begin{array}{c} \mathbf{T_0}\ \mathbf{T_1}\ \cdots\cdots\cdots\ \mathbf{T_8} \end{array}$$

$$\mathbf{A} = \begin{bmatrix} 2 & -1 & 0 & 0 & 0 & 0 & 0 & 0 & 0 \\ -1 & 4 & -1 & 0 & 0 & 0 & 0 & 0 & 0 \\ 0 & -1 & 4 & -1 & 0 & 0 & 0 & 0 & 0 \\ 0 & 0 & -1 & 4 & -1 & 0 & 0 & 0 & 0 \\ 0 & 0 & 0 & -1 & 4 & -1 & 0 & 0 & 0 \\ 0 & 0 & 0 & 0 & -1 & 4 & -1 & 0 & 0 \\ 0 & 0 & 0 & 0 & 0 & -1 & 4 & -1 & 0 \\ 0 & 0 & 0 & 0 & 0 & 0 & -1 & 4 & -1 \\ 0 & 0 & 0 & 0 & 0 & 0 & 0 & -1 & 4 \end{bmatrix} \begin{array}{l} \text{node 0} \\ \text{node 1} \\ \vdots \\ \\ \\ \\ \\ \\ \text{node 9} \end{array}$$

Note that one of the coefficients for node 9 does not appear. This omission arises because the temperature of the adjacent node (9 + 1) is assumed to remain constant and equal to the initial temperature T_0. This assumption is incorporated into the coefficient matrix **C**, which appears on the right-hand side of Eqs. (P12.2) and (P12.3):

$$\mathbf{C} = \begin{bmatrix} 56.1 + T_0^p \\ 2 \cdot T_1^p \\ 2 \cdot T_2^p \\ 2 \cdot T_3^p \\ 2 \cdot T_4^p \\ 2 \cdot T_5^p \\ 2 \cdot T_6^p \\ 2 \cdot T_7^p \\ 2 \cdot T_8^p + T_0 \end{bmatrix} \begin{array}{l} \text{node 0} \\ \text{node 1} \\ \vdots \\ \\ \\ \\ \\ \\ \text{node 9} \end{array}$$

To obtain the nodal temperatures at step $p + 1$, the inverse of the coefficient matrix, $\mathbf{A}^{-1}$, is first computed. This matrix is then multiplied by the column vector **C**, which depends on the nodal temperatures at step *p*. After computing the new temperatures, the vector **C** is updated and the multiplication by $\mathbf{A}^{-1}$ is repeated.

In the present case, this procedure is carried out five times, corresponding to a time increment of $\Delta t = 24$ s, for a total elapsed time of 120 s. Upon completion of the recursion, the temperature at node 2 is found to be 44.2 °C. The temperature profiles for all nodes at the discrete time steps are shown in Fig. P12.3.

The same profile displayed in a 3D plot is shown in Fig. P12.4.

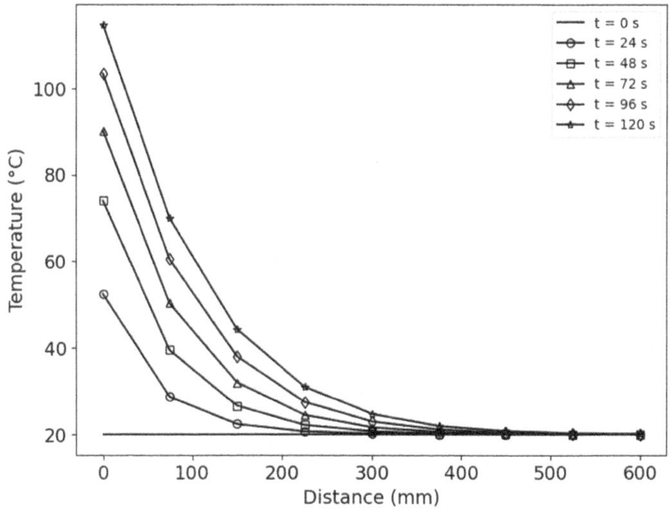

Fig. P12.3 Temperature profiles as a function of time and node position

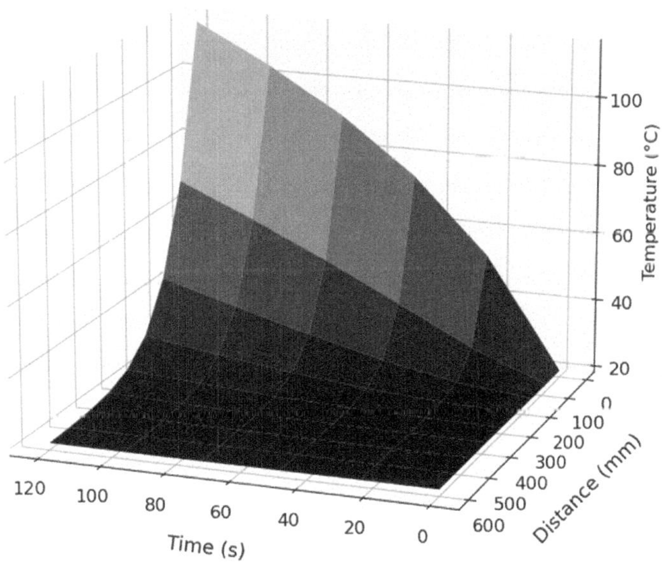

Fig. P12.4 Temperature profiles as a function of time and node position

The Python code developed for this problem is:

```
import numpy as np
import matplotlib.pyplot as plt

# Construct matrix A and its inverse
A = np.diag(4*np.ones(9)) + np.diag(-1*np.ones(8), 1) + np.diag(-1*np.ones(8), -1)
A[0,0] = 2
Ainv = np.linalg.inv(A)

# Initial condition
C = np.array([76.1, 40, 40, 40, 40, 40, 40, 40, 60])

# Time recursion (5 iterations)
D = np.zeros((9, 5))
for i in range(5):
    D[:, i] = np.dot(Ainv, C)
    C = np.empty(9)
    C[0] = 56.1 + D[0, i]
    C[1:8] = 2*D[1:8, i]
    C[8] = 2*D[8, i] + 20

# FIGURES
# Prepare axis data
pos = np.arange(0, 601, 75)  # mm
tpo = np.arange(0, 121, 24)  # s
C0 = 20*np.ones(9)

plt.figure(1, figsize=(8, 6))
plt.plot(pos, C0, 'k', label='t = 0 s')
markers = ['o', 's', '^', 'd', '*']
for i in range(5):
    plt.plot(pos, D[:, i], color='k', marker=markers[i], fillstyle='none',
             label=f't = {24*(i+1)} s')
plt.xlabel('Distance (mm)', fontsize=14)
plt.ylabel('Temperature (°C)', fontsize=14)
plt.tick_params(axis='both', labelsize=14)
plt.legend(fontsize=10)

# Figures 2 and 3
X, Y = np.meshgrid(tpo, pos)
Z = np.column_stack((C0, D))
print(Z.shape)

# 2D plot: Heat map
plt.figure(2)
plt.pcolormesh(X, Y, Z, cmap='plasma',
               vmin=np.nanmin(Z), vmax=np.nanmax(Z))
plt.colorbar(label='Temperature (°C)')
plt.xlabel('Time (s)')
plt.ylabel('Distance (mm)')

# 3D plot
fig = plt.figure(3, figsize=(7, 7))
ax = fig.add_subplot(111, projection='3d')
surf = ax.plot_surface(X, Y, Z, cmap='gray',
                       vmin=np.nanmin(Z), vmax=np.nanmax(Z))
ax.set_xlabel('Time (s)', fontsize=13, labelpad=12)
ax.set_ylabel('Distance (mm)', fontsize=13, labelpad=12)
ax.set_zlabel('Temperature (°C)', fontsize=12)
ax.view_init(azim=110, elev=15)
plt.tick_params(axis='both', labelsize=12)
fig.tight_layout()

plt.show()
```

The reader is encouraged to solve this problem using the method of lines.

6.13 Problem 13 Method of Lines Applied to Partial Differential Equations (PDEs): Mixing and Cooling

The mixing and cooling process in a perfectly stirred reactor is carried out under conditions that ensure complete homogenization of the components of the liquid mixture. The fluid is contained in a steel reactor, whose walls have thickness d and thermal conductivity k. Heat transfer between the fluid and the inner wall (or inner face) of the reactor, of area A, occurs by internal convection (h_i). On the other hand, heat losses to the surroundings at temperature (Ta) occur by external convection (h_o) and thermal radiation ($T_{sur} = T_a$ and thermal emissivity ε). It may be assumed that heat losses through the reactor heads are negligible, and that the external area of the reactor wall is approximately A. If the fluid and the reactor steel are initially at 150 °C and 30 °C, respectively:

(a) Plot the temperature profiles in the wall (for $0 \leq x \leq d$, using 20 internal nodes) as a function of time and position, together with the thermal dynamics of the fluid inside the reactor, for a total simulation time of 2 h.
(b) After 1000 s of operation, what is the temperature difference between the inner face ($x = 0$) and the outer face ($x = d$) of the reactor wall?

Data: Mass of the fluid, $m = 800$ kg; fluid heat capacity, $c_p = 4160\,\text{J kg}^{-1}\,\text{K}^{-1}$; wall thickness, $d = 0.08$ m; heat-transfer area, $A = 3$ m^2; thermal diffusivity of steel, $\alpha = 3.95\,10^{-6}$ m^2/s; inner heat-transfer coefficient, $h_i = 1200\,\text{W m}^{-2}\,\text{K}^{-1}$; outer heat-transfer coefficient, $h_o = 10\,\text{W m}^{-2}\,\text{K}^{-1}$; ambient temperature, $T_a = 15$ °C; thermal conductivity of steel, $k = 15\,\text{W m}^{-1}\,\text{K}^{-1}$; thermal emissivity, $\varepsilon = 0.98$.

Solution

This problem presents several challenges. On the one hand, because the fluid is assumed to be perfectly mixed, its temperature may be modeled by a lumped energy balance:

$$m \cdot cp \frac{\mathrm{d}T_f}{\mathrm{d}t} = -h_i \cdot A \cdot (T_f - T(x = 0)), \tag{P13.1}$$

where T_f denotes the fluid temperature and $T(x = 0)$ represents the wall temperature at the fluid–wall interface.

On the other hand, because the wall is relatively thick and has a low thermal conductivity, its temperature varies with the spatial coordinate x. Furthermore, since the heat-transfer area A is approximately the same on both sides of the reactor, the wall temperature T may be described using Fourier's heat conduction equation. The formulation adopted here assumes constant cross-sectional area and one-dimensional heat flow in the x-direction:

$$\frac{\partial T}{\partial t} = \alpha \frac{\partial^2 T}{\partial x^2} \tag{P13.2}$$

This problem involves two boundary conditions. The first applies at $x = 0$, where the fluid mixture is in contact with the reactor wall. At this location, the heat transferred from the fluid by convection is conducted into the interior of the wall. Accordingly,

$$h_i \cdot (T_\mathrm{f} - T_{x=0}) = -k \left.\frac{\mathrm{d}T}{\mathrm{d}x}\right|_{x=0}. \tag{P13.3}$$

The second boundary condition applies at the opposite side of the wall, $x = \mathrm{d}$, where the heat conducted through the wall is transferred to the surroundings by convection and thermal radiation:

$$-k\left.\frac{\mathrm{d}T}{\mathrm{d}x}\right|_{x=d} = h_o \cdot (T_{x=d} - T_a) + \varepsilon \cdot \sigma \cdot \left(T_{x=d}^4 - T_a^4\right). \tag{P13.4}$$

Given this configuration of the problem, the problem consists of an ordinary differential equation describing the fluid temperature (Eq. (P13.1)) coupled, through the boundary condition at $x = 0$ (Eq. (P13.3)), to a partial differential equation governing heat conduction in the wall (Eq. (P13.2)). When the second boundary condition (Eq. (P13.4)) is included, the system becomes fully coupled, linking the fluid temperature T_f to the ambient temperature T_a.

To solve this coupled system, the method of lines is employed. This approach discretizes the spatial derivatives using finite differences while retaining ordinary differential equations for the time derivatives. The resulting system of ODEs is then integrated using a suitable numerical integrator, such as `solve_ivp` from the SciPy library.

The reactor wall is discretized into n interior nodes separated by equal distances. Boundary conditions (P13.3) and (P13.4) are used to evaluate the temperatures at $T(x = 0)$ and $T(x = d)$, respectively, which are then incorporated into the finite-difference approximations for nodes 1 and n. The following schematic illustrates the spatial discretization of the wall (Fig. P13.1).

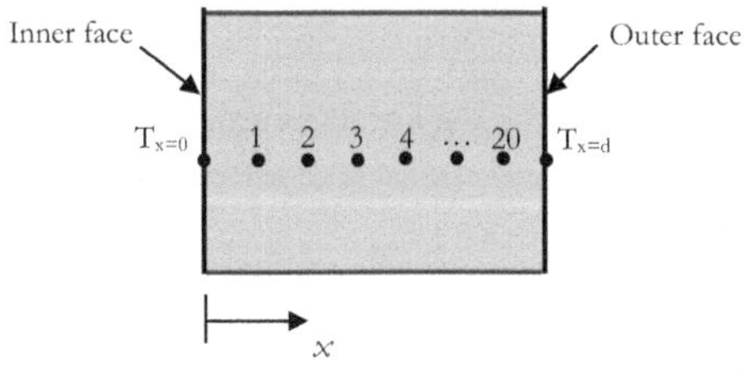

Fig. P13.1 Discretization of the wall using 20 interior nodes

To discretize the boundary condition at the inner surface (Eq. (P13.4)), a second-order forward finite-difference approximation is applied to the first derivative. Solving for the temperature at the inner surface, $T(x = 0)$, yields:

$$T_{x=0} = \frac{2\Delta x \cdot h_i \cdot T_f + 4k \cdot T_1 - k \cdot T_2}{2\Delta x \cdot h_i + 3k} \quad \text{(P13.5)}$$

This temperature is required when evaluating the second derivative in Fourier's equation for node 1, through the term T_{i-1}. To discretize the spatial derivative in Eq. (P13.3), a second-order central finite-difference approximation is employed for the second derivative, yielding:

$$\frac{\partial T_i}{\partial t} = \alpha\left(\frac{T_{i+1} - 2T_i + T_{i-1}}{\Delta x^2}\right). \quad \text{(P13.6)}$$

It should be noted that only the spatial derivatives are discretized. The temporal derivatives are handled by the numerical integrator, and for each node i ($i = 1 \ldots n$) an ordinary differential equation is obtained.

The second boundary condition (Eq. (P13.4)) is used to determine the temperature at $x = d$, which is required for the evaluation of T_{n+1} at the last interior node. The challenge lies in solving for $T(x = d)$, since the radiative heat-transfer term introduces a fourth-power dependence on temperature, while the discretized spatial derivative contributes a first-order term.

To address this difficulty, the spatial derivative is discretized using a second-order backward finite-difference approximation for the first derivative. Then, at each integration step, the nonlinear algebraic equation for $T(x = d)$ is solved numerically using the `fsolve` routine. This procedure will become clearer upon examination of the corresponding implementation. The resulting discretized boundary condition is given below:

$$k\left[\frac{3T_{x=d} - 4T_n + T_{n-1}}{2\Delta x}\right] = h_o \cdot (T_{x=d} - T_a) + \varepsilon \cdot \sigma \cdot \left(T_{x=d}^4 - T_a^4\right). \quad \text{(P13.7)}$$

Thus, the code developed for this problem is:

```
import numpy as np
from scipy.integrate import solve_ivp
from scipy.optimize import fsolve
import matplotlib.pyplot as plt

# Discretization
d = 0.08
n = 20
x = np.linspace(0, d, n+2)  # 20 interior nodes + 2 boundaries
dx = x[1] - x[0]
x = x[1:-1]

# General parameters
ho = 10           # W/m2K
hi = 1200         # W/m2K
Ta = 15 + 273.15  # K
k  = 15           # W/mK

# Reactor parameters
A = 3             # m2
m = 800           # kg
cpl = 4160        # J/kg*K
alpha = 3.95e-6 # AISI 304
epsilon = 0.98

# Balance at the outer interface: conduction = convection + radiation
def ecrad(T, Tn1, Tn2):
    # Second-order backward finite difference for the first derivative
    return -k*(3*T - 4*Tn1 + Tn2)/(2*dx) - ho*(T - Ta) - epsilon*5.67e-8*(T**4 - Ta**4)

# Differential model
def modelo(t, T):
    # Initialize nodes: n + 1 (internal wall nodes + fluid)
    dT = np.empty(n+1)

    # First wall node: convection = conduction
    Tbi = (2*dx*hi*T[0] + 4*k*T[2] - k*T[3])/(2*dx*hi + 3*k)

    # Last wall node: conduction = convection + radiation
    [Tbd] = fsolve(ecrad, 350, args=(T[-1], T[-2]), xtol=1e-9)

    # ODE for the fluid
    dT[0] = -hi*A*(T[0] - Tbi)/(m*cpl)

    # ODEs for interior wall nodes
    for ec in range(1, n+1):
        if 1 < ec < n:
            # Central difference for the second derivative
            dT[ec] = alpha*(T[ec+1] - 2*T[ec] + T[ec-1])/dx**2
        elif ec == 1:
            # First interior node uses boundary condition Tbi
            dT[ec] = alpha*(T[ec+1] - 2*T[ec] + Tbi)/dx**2
        elif ec == n:
            # Last interior node uses boundary condition Tbd
            dT[ec] = alpha*(Tbd - 2*T[ec] + T[ec-1])/dx**2
    return dT
```

```
# Model integration
Twall0 = 30     # °C
Tfluid0 = 150   # °C
T0 = np.array([Tfluid0] + [Twall0]*n) + 273.15
sol = solve_ivp(modelo, (0, 3600*2), T0, method='BDF', rtol=1e-9, atol=1e-9)
T = sol.y

# Dimensionality
print(f"The solution t has {sol.t.shape[0]} rows.")
print(f"The solution T has {T.shape[0]} rows (nodes) and {T.shape[1]} columns (time).")

# Compute fluid-wall (Tci) and wall-air (Tcd) interface temperatures
Tci = (2*dx*hi*T[0] + 4*k*T[2] - k*T[3])/(2*dx*hi + 3*k)  # inner face
Tcd = np.empty(T.shape[1])
for dt in range(T.shape[1]):
    [Tcd[dt]] = fsolve(ecrad, 310, args=(T[-1,dt], T[-2,dt]), xtol=1e-9)  # outer face

# Part (b): Interpolate interface temperatures at t = 1000 s
Tci_1000 = np.interp(1000, sol.t, Tci)   # inner face
Tcd_1000 = np.interp(1000, sol.t, Tcd)   # outer face
print("The temperature difference between both faces is",
      round(Tci_1000 - Tcd_1000, 1), "°C.")

# Figures
# Figure 1: temperature profiles for fluid + wall nodes
T = T - 273.15
Tf = T[0]        # fluid temperature
T = T[1:, :]     # internal wall nodes

fig = plt.figure(figsize=(10, 8))
plt.plot(sol.t, Tf, color='k', ls='-.', lw=2, label='Fluid temperature')
plt.plot(sol.t, T.T, color='k', lw=0.5)
plt.plot(sol.t, Tci - 273.15, color='k', lw=1.5, label='Inner face')
plt.plot(sol.t, Tcd - 273.15, color='k', alpha=0.7, lw=3.5, label='Outer face')
plt.xlabel('Time (s)', fontsize=14)
plt.ylabel('Temperature (°C)', fontsize=14)
plt.tick_params(axis='both', labelsize=14)
plt.legend(fontsize=10)

# Figure 2: 3D temperature distribution in the wall
Tm, X = np.meshgrid(sol.t, x)

fig = plt.figure(figsize=(8, 10))
ax = fig.add_subplot(111, projection='3d')
surf = ax.plot_surface(Tm, X, T, cmap='viridis', edgecolor='none')
ax.set_xlabel('Time (s)', fontsize=14, labelpad=10)
ax.set_ylabel('Wall (m)', fontsize=14, labelpad=10)
ax.set_zlabel('Temperature (°C)', fontsize=14)
plt.tick_params(axis='both', labelsize=12)
ax.view_init(azim=290, elev=20)
fig.tight_layout()

plt.show()
```

Figure P13.2 shows the temperature profiles of the fluid and the reactor wall during the first two hours of simulation.

Note that, if the simulation is extended over a sufficiently long-time horizon, all temperatures converge toward the ambient temperature T_a, eventually reaching thermal equilibrium (Fig. P13.3).

Figure P13.4 shows the temperature profiles of the internal wall nodes during the first two hours of simulation.

To address part (b), the temperatures at both wall faces are interpolated at $t = 1000$ s, and their difference is then computed, yielding:

$$\Delta\ \textbf{Temperature}\ \text{between both faces} = 25.6\ ^{\circ}\text{C}$$

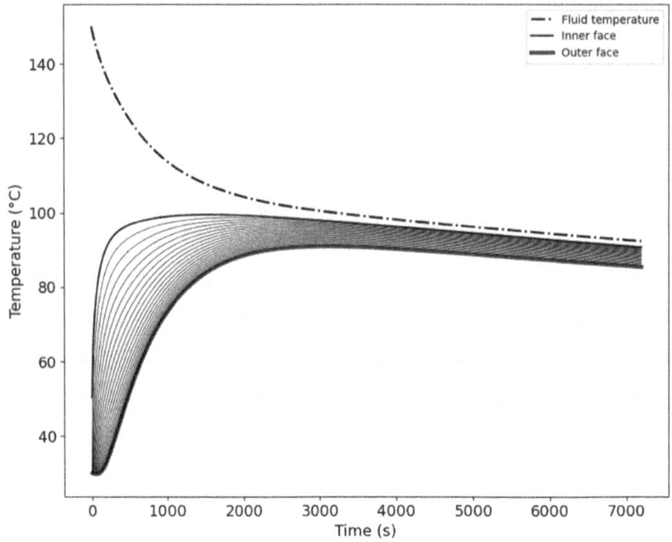

Fig. P13.2 Temperature profiles as functions of time for the fluid, both faces of the reactor wall, and the internal wall nodes

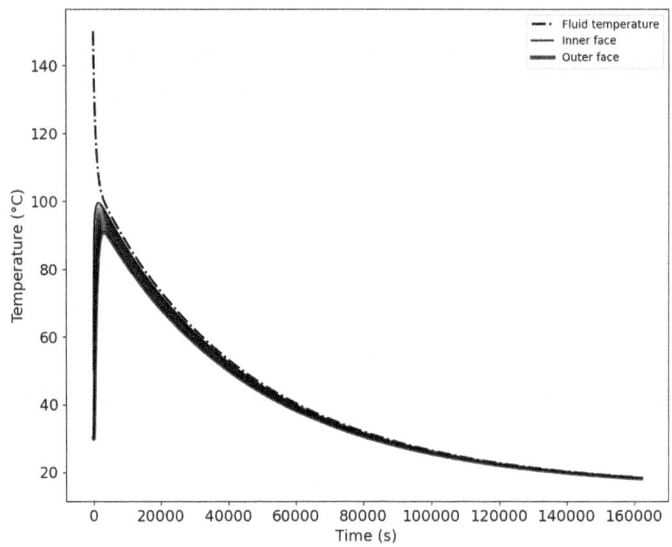

Fig. P13.3 Temperature profiles of the system at $t = 45$ h

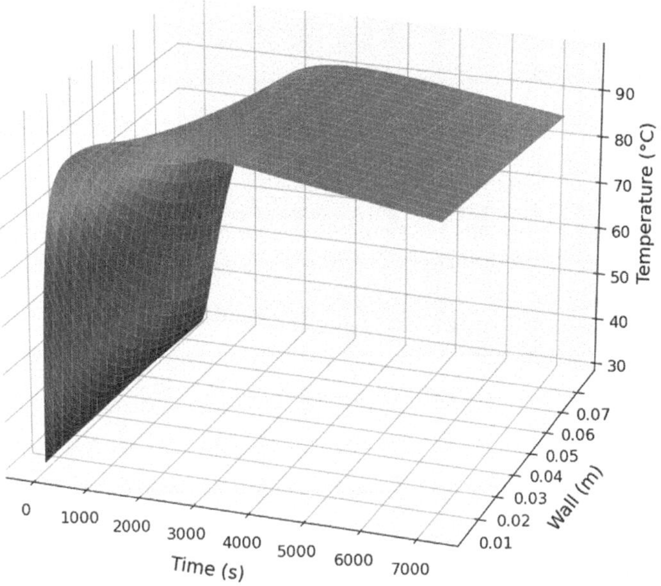

Fig. P13.4 Temperature profiles of the internal wall nodes of the reactor for a simulation time of t = 2 h

6.14 Problem 14 Method of Lines Applied to Partial Differential Equations (PDEs): Radial Dispersion and Wall Reaction in a Cylindrical Reactor

Consider a cylindrical reactor (length L and radius R) in which radial dispersion of a species occurs, and the species reacts at the inner wall of the cylinder. Assuming a uniform inlet fluid velocity, the mass balance for the compound reacting at the wall is given by:

$$v_0 \frac{\partial C}{\partial z} = D_{AB}\left(\frac{1}{r}\frac{\partial}{\partial r}\left(r\frac{\partial C}{\partial r}\right)\right), \tag{P14.1}$$

where v_0 is the bulk flow velocity and D_{AB} is the radial dispersion coefficient within the reactor. The governing equation is subject to the following conditions:

At $r = 0$ (centerline):

$$\frac{\partial C}{\partial r} = 0. \tag{P14.2}$$

At $r = R$ (wall):

$$-D_{AB}\left(\frac{\partial C}{\partial r}\right) = kC. \tag{P14.3}$$

At z = 0:

$$C = C_0. \tag{P14.4}$$

Apply the method of lines to calculate the concentration C as a function of the axial (z) and radial (r) coordinates. Use 20 radial nodes, including $r = 0$ and $r = R$. Set the following parameter values: $C_0 = 5$ mol/m^3, $D_{AB} = 1.2 \times 10^{-4}$ m^2/s, $R =$ 0.1 m, $L = 1$ m, $v_0 = 0.05$ m/s, $k = 1.2$ m/s. Plot the resulting concentration profiles as functions of z and r.

Now consider a parabolic velocity profile, given by:

$$v(r) = 2v_0\left[1 - \left(\frac{r}{R}\right)^2\right]. \tag{P14.5}$$

Plot the resulting concentration profiles using the same number of radial nodes. In addition, at $r = 0$ and $r = R$, compare them with the profiles obtained under the plug-flow assumption, i.e., $v(r) = v_0$.

Solution

To solve the system, we apply the method of lines, i.e., we discretize the spatial derivatives (here, the radial derivatives) using finite differences and then integrate the resulting system of ordinary differential equations along the remaining independent variable, z. As in the previous problem, this approach ultimately requires an ODE integrator such as `solve_ivp` from the Scipy library.

We discretize the interval $0 \leq r \leq R$ using $n = 20$ radial nodes, including both boundaries $r = 0$ and $r = R$. The radial coordinate is defined as:

$$r_i = i\Delta r, \quad i = 0, 1, \ldots, n-1 \quad \Delta r = \frac{R}{n-1}, \tag{P14.6}$$

where the node indexing starts at $i = 0$. Consequently, the last node corresponds to $i = n - 1$, rather than $i = n$, which is consistent with Python indexing conventions.

The symmetry condition (Eq. (P14.2)) can be expressed using a second-order finite-difference approximation in terms of the interior nodes:

$$C_0 = \frac{4C_1 - C_2}{3}. \tag{P14.7}$$

This avoids evaluating the PDE at $r = 0$, where the $1/r$ term would be singular.

At the wall ($r = R$), the boundary condition in Eq. (P14.3), is discretized using a second-order backward finite-difference approximation for the radial derivative:

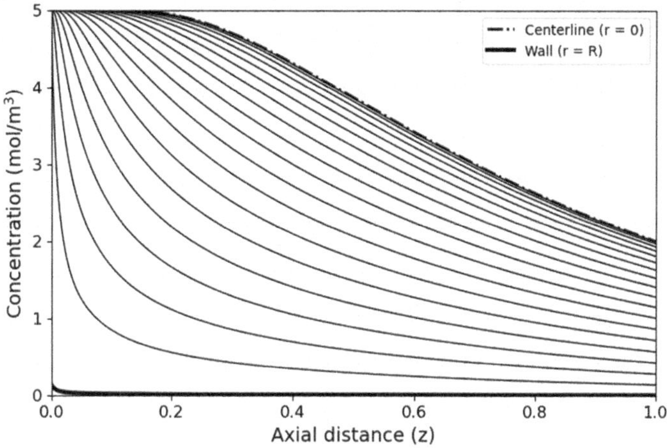

Fig. P14.1 Concentration profiles (uniform velocity), $n = 20$ nodes

$$\left.\frac{\mathrm{d}C}{\mathrm{d}r}\right|_{r=R} = \frac{3C_{n-1} - 4C_{n-2} + C_{n-3}}{2\Delta r}. \tag{P14.8}$$

Now, substituting Eq. (P14.9) into Eq. (P14.4), we obtain the explicit expression for the boundary node C_{n-1}:

$$C_{n-1} = \frac{4C_{n-2} - C_{n-3}}{3 + 2k\Delta r/D_{AB}}. \tag{P14.9}$$

The radial operator of the PDE (Eq. (P14.2)) can be written as:

$$\frac{1}{r}\frac{\partial}{\partial r}\left(r\frac{\partial C}{\partial r}\right) = \frac{\partial^2 C}{\partial r^2} + \frac{1}{r}\frac{\partial C}{\partial r}. \tag{P14.10}$$

Now, for each interior node $i = 1, \ldots, n - 2$ (so that $r_i > 0$), we use second-order central differences:

$$\left.\frac{\partial C}{\partial r}\right|_i = \frac{C_{i+1} - C_{i-1}}{2\Delta r} \qquad \left.\frac{\partial^2 C}{\partial r^2}\right|_i = \frac{C_{i+1} - 2C_i + C_{i-1}}{\Delta r^2}. \tag{P14.11}$$

Substituting into Eq. (P14.1) yields the method-of-lines system:

$$\frac{\mathrm{d}C_i}{\mathrm{d}z} = \frac{D_{AB}}{v(r_i)}\left[\frac{C_{i+1} - C_{i-1}}{2r_i\Delta r} + \frac{C_{i+1} - 2C_i + C_{i-1}}{\Delta r^2}\right], \quad i = 1, \ldots, n-2 \tag{P14.12}$$

The inlet condition (Eq. (P14.4)) provides the initial condition for integration in z: $C_i(z = 0) = C_0$ for all radial nodes.

Note that, in Eq. (P14.12), the velocity term v can be directly replaced by either the plug-flow assumption ($v = v_0$) or a parabolic profile (Eq. (P14.5)). In practice, the model is solved twice, once for plug flow and once for the parabolic profile, and the resulting concentration fields are compared.

Figure P14.1 presents the concentration profiles for the plug-flow case, whereas Fig. P14.2 shows the corresponding profiles obtained with the parabolic velocity distribution.

Figure P14.3 compares the concentration profiles obtained with plug flow and with the parabolic velocity distribution. Note that at the wall ($r = R$), both curves overlap for the plug-flow and parabolic cases.

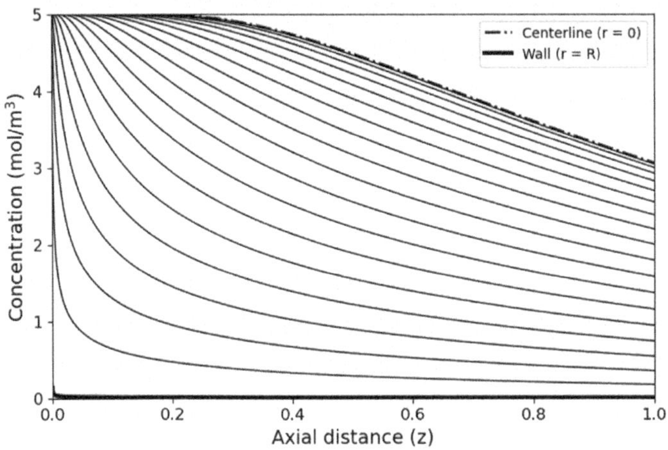

Fig. P14.2 Concentration profiles (parabolic velocity), $n = 20$ nodes

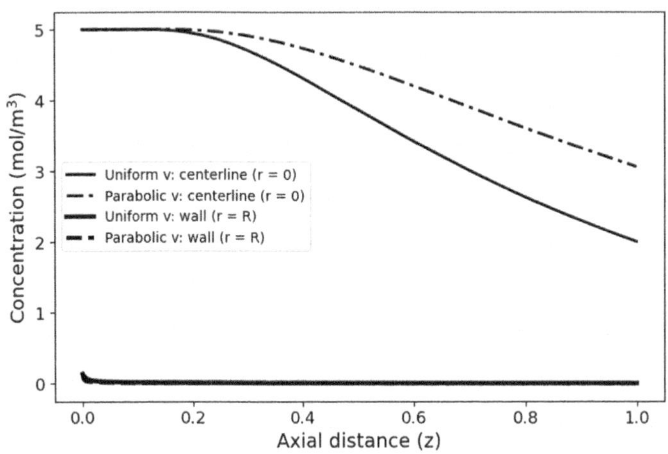

Fig. P14.3 Comparison of concentration profiles at $r = 0$ and $r = R$ ($n = 20$ nodes)

The Python code developed for this problem is:

```
import numpy as np
import matplotlib.pyplot as plt
from scipy.integrate import solve_ivp

# -----------------------------
# Discretization
# -----------------------------
L = 1.0
R = 0.1
n = 20  # radial nodes including r=0 and r=R

r = np.linspace(0.0, R, n)
dr = r[1] - r[0]

# Interior radial nodes (exclude r=0 and r=R)
r_int = r[1:-1]              # size = n-2
n_int = r_int.size           # n_int = n-2

# -----------------------------
# Physical properties
# -----------------------------
v0 = 0.05
Dab = 1.2e-4
k = 1.2

# -----------------------------
# Initial condition
# -----------------------------
C0 = 5.0  # mol/m^3
y0 = C0 * np.ones(n_int)      # state contains ONLY interior nodes

print(f"n = {n}, dr = {dr:.6f}, interior state size = {y0.size}")

def reconstruct_full_C_from_interior(C_int, n, dr, Dab, k):
    """
    Reconstruct full C(z,r) including r=0 and r=R from interior nodes only.

    Parameters
    ----------
    C_int : array (m, n-2)
        Solution at interior nodes r[1]..r[n-2].
    Returns
    -------
    C_full : array (m, n)
        Reconstructed solution at all nodes r[0]..r[n-1].
    """
    m = C_int.shape[0]
    C_full = np.empty((m, n), dtype=float)

    # Fill interior nodes
    C_full[:, 1:-1] = C_int

    # Boundary at r=0 (second-order symmetry): C0 = (4*C1 - C2)/3
    C_full[:, 0] = (4.0 * C_full[:, 1] - C_full[:, 2]) / 3.0

    # Boundary at r=R (second-order Robin):
    # C_R = (4*C_{n-2} - C_{n-3}) / (3 + 2*k*dr/Dab)
    denom = 3.0 + 2.0 * k * dr / Dab
    C_full[:, -1] = (4.0 * C_full[:, -2] - C_full[:, -3]) / denom

    return C_full

def build_full_vector_from_state(y, n, dr, Dab, k):
    """
    Build a full vector C[0..n-1] from the interior-state vector y (length n-2),
    using second-order boundary reconstructions at r=0 and r=R.
    """
```

```
    C = np.empty(n, dtype=float)

    # Fill interior
    C[1:-1] = y

    # r=0 boundary (second-order symmetry)
    C[0] = (4.0 * C[1] - C[2]) / 3.0

    # r=R boundary (second-order Robin)
    denom = 3.0 + 2.0 * k * dr / Dab
    C[-1] = (4.0 * C[-2] - C[-3]) / denom

    return C

def modelo(z, y, n, r_int, dr, v0, Dab, k, R):
    """
    Part (a): v(r) = v0 (uniform / plug flow).
    State y has length n-2 corresponding to r[1]..r[n-2].
    """
    C = build_full_vector_from_state(y, n, dr, Dab, k)

    dy = np.zeros_like(y)

    # Interior indices in the full vector correspond to i = 1..n-2
    # State index j corresponds to full index i = j+1
    for j in range(y.size):
        i = j + 1
        term1 = (C[i + 1] - C[i - 1]) / (r[i] * 2.0 * dr)
        term2 = (C[i + 1] - 2.0 * C[i] + C[i - 1]) / (dr ** 2)
        dy[j] = (Dab / v0) * (term1 + term2)

    return dy

def modelo2(z, y, n, r_int, dr, v0, Dab, k, R):
    """
    Part (b): v(r) = 2*v0*(1-(r/R)^2) (parabolic profile).
    State y has length n-2 corresponding to r[1]..r[n-2].
    """
    C = build_full_vector_from_state(y, n, dr, Dab, k)

    dy = np.zeros_like(y)

    for j in range(y.size):
        i = j + 1
        v_i = 2.0 * v0 * (1.0 - (r[i] / R) ** 2)
        term1 = (C[i + 1] - C[i - 1]) / (r[i] * 2.0 * dr)
        term2 = (C[i + 1] - 2.0 * C[i] + C[i - 1]) / (dr ** 2)
        dy[j] = (Dab / v_i) * (term1 + term2)

    return dy

# ------------------------------
# Integration: Part (a)
# ------------------------------
sol_a = solve_ivp(
    fun=lambda z, y: modelo(z, y, n, r_int, dr, v0, Dab, k, R),
    t_span=(0.0, L),
    y0=y0,
    method="BDF",
    rtol=1e-9,
    atol=1e-12
)

z_a = sol_a.t
C_a_full = reconstruct_full_C_from_interior(sol_a.y.T, n, dr, Dab, k)

print("Part (a) integration complete.")
print("z points:", z_a.size, "| C_a_full shape:", C_a_full.shape)

# ------------------------------
# Integration: Part (b)
```

```
# -----------------------------
sol_b = solve_ivp(
    fun=lambda z, y: modelo2(z, y, n, r_int, dr, v0, Dab, k, R),
    t_span=(0.0, L),
    y0=y0,
    method="BDF",
    rtol=1e-9,
    atol=1e-12
)

z_b = sol_b.t
C_b_full = reconstruct_full_C_from_interior(sol_b.y.T, n, dr, Dab, k)

print("Part (b) integration complete.")
print("z points:", z_b.size, "| C_b_full shape:", C_b_full.shape)

# ----------------
# 2D plots
# ----------------
# (a) Uniform velocity
plt.figure(figsize=(8, 5))

# All interior radial nodes (thin black, no legend entry)
plt.plot(z_a, C_a_full[:, 1:-1], color="black", lw=1.0)

# Centerline r=0 (dash-dot, legend entry)
plt.plot(z_a, C_a_full[:, 0], color="black", lw=1.8, ls="-.",
         label="Centerline (r = 0)")

# Wall r=R (thick solid, legend entry)
plt.plot(z_a, C_a_full[:, -1], color="black", lw=3.0, ls="-",
         label="Wall (r = R)")

plt.axis([0, 1, 0, 5])
plt.xlabel("Axial distance (z)", fontsize=14)
plt.ylabel("Concentration (mol/m$^3$)", fontsize=14)
plt.tick_params(axis='both', labelsize=12)
# plt.title(f"(a) Concentration profiles (uniform velocity), n = {n}")
plt.legend()
plt.show()

# (b) Parabolic velocity
plt.figure(figsize=(8, 5))

plt.plot(z_b, C_b_full[:, 1:-1], color="black", lw=1.0)
plt.plot(z_b, C_b_full[:, 0], color="black", lw=1.8, ls="-.",
         label="Centerline (r = 0)")
plt.plot(z_b, C_b_full[:, -1], color="black", lw=3.0, ls="-",
         label="Wall (r = R)")

plt.axis([0, 1, 0, 5])
plt.xlabel("Axial distance (z)", fontsize=14)
plt.ylabel("Concentration (mol/m$^3$)", fontsize=14)
plt.tick_params(axis='both', labelsize=12)
# plt.title(f"(b) Concentration profiles (parabolic velocity), n = {n}")
plt.legend()

plt.show()

# -----------------------------
# Comparison at r=0 and r=R
# -----------------------------
i_center = 0
i_wall = n - 1

plt.figure(figsize=(8, 5))

# Centerline
plt.plot(z_a, C_a_full[:, i_center], color="black", lw=2.0, ls="-",
         label="Uniform v: centerline (r = 0)")
plt.plot(z_b, C_b_full[:, i_center], color="black", lw=2.0, ls="-",
```

```
        dashes=(6, 2, 1.5, 2), label="Parabolic v: centerline (r = 0)")

# Wall
plt.plot(z_a, C_a_full[:, i_wall], color="black", lw=3.2, ls="-",
         label="Uniform v: wall (r = R)")
plt.plot(z_b, C_b_full[:, i_wall], color="black", lw=3.0, ls="--",
         label="Parabolic v: wall (r = R)")

plt.xlabel("Axial distance (z)", fontsize=14)
plt.ylabel("Concentration (mol/m$^3$)", fontsize=14)
plt.tick_params(axis='both', labelsize=12)
# plt.title("Comparison of concentration profiles at r = 0 and r = R")
plt.legend(loc='center left')

plt.show()
```

References

Bailey, J. E. (1986). *Biochemical engineering fundamentals*. McGraw-Hill Chemical Engineering Series.

Bequette, B. W. (1998). *Process dynamics: Modeling, analysis and simulation*. Prentice Hall.

Bird, R. B., Stewart, W. E., & Lightfoot, E. N. (2007). *Transport phenomena* (2nd ed.). John Wiley & Sons.

Englezos, P., & Kalogerakis, N. (2001). *Applied parameter estimation for chemical engineers*. Marcel Dekker.

Incropera, F. P., DeWitt, D. P., Bergman, T. L., & Lavine, A. S. (2007). *Fundamentals of heat and mass transfer* (6th ed.). John Wiley & Sons.

Gelmi, C., Pérez-Correa, R., & Agosin, E. (2002). Modelling *Gibberella fujikuroi* growth and GA3 production on solid-state fermentation. *Process Biochemistry, 37*(9), 1033–1040.

Sánchez, B. J., Soto, D. C., Jorquera, H., Gelmi, C., & Pérez-Correa, J. R. (2014). HIPPO: An iterative reparametrization method for identification and calibration of dynamic bioreactor models of complex processes. *Industrial and Engineering Chemistry Research, 53*(48), 18514–18525.

Luyben, W. L. (1973). *Process modeling, simulation, and control for chemical engineers*. McGraw-Hill Chemical Engineering Series.

Appendix

Python Resources

Below is a list that is not intended to be exhaustive, but rather illustrative of useful libraries and tools for further exploration.

GitHub: https://github.com/

A collaborative development platform for hosting and reviewing code, managing projects, and building software with other contributors.

Google Colab (Colaboratory): https://colab.research.google.com/

A free, cloud-based Jupyter Notebook environment provided by Google. It lets you run Python code directly in the browser without installing anything and offers easy access to GPUs/TPUs for accelerated computing. Ideal for quick experimentation, teaching, sharing notebooks via links, and collaborating in real time.

Jupyter: https://jupyter.org/

Provides Jupyter Notebooks, a powerful tool for creating documents that combine executable code, explanatory text, equations, and visualizations.

Matplotlib: https://matplotlib.org/

A plotting library for creating static, animated, and interactive visualizations in Python. Very useful for presenting numerical analysis results.

NumPy: https://numpy.org/

A fundamental library for scientific computing in Python. It provides support for large, multidimensional arrays and matrices, along with a broad collection of mathematical functions.

Pandas: https://pandas.pydata.org/

H. Jorquera and C. A. Gelmi, *Numerical Methods in Chemical Process Engineering Using Python*, https://doi.org/10.1007/978-3-032-22958-8

Essential for data analysis in Python, it provides flexible data structures and tools to work efficiently with structured data.

PyPI (Python Package Index): https://pypi.org/

The official repository for software packages developed and shared by the Python community. Ideal for discovering and sharing useful libraries. An excellent starting point for any need.

Scikit-learn: https://scikit-learn.org/

An efficient library for machine learning and data analysis in Python. It includes tools for classification, regression, clustering, and dimensionality reduction.

SciPy: https://scipy.org/

Built on top of NumPy, SciPy facilitates numerical computation, optimization, integration, interpolation, statistics, Fourier transforms, signal processing, and much more.

Statsmodels: https://www.statsmodels.org/

A Python library that provides classes and functions for estimating different statistical models, as well as performing statistical tests and data exploration.

SymPy: https://www.sympy.org/

A Python library for symbolic mathematics. It enables exact algebraic computations.